ASCE STANDARDS

In 2014, the Board of Direction approved revisions to the ASCE Rules for Standards Committees to govern the writing and maintenance of standards developed by ASCE. All such standards are developed by a consensus standards process managed by the ASCE Codes and Standards Committee (CSC). The consensus process includes balloting by a balanced standards committee, and reviewing during a public comment period. All standards are updated or reaffirmed by the same process. Extensions beyond a five-year period shall require the approval of both the appropriate Executive Committee and CSC. Requests for formal interpretations shall be processed in accordance with Section 7 of ASCE Rules for Standards Committees, which are available at www.asce.org. Errata, addenda, supplements, and interpretations, if any, for this standard can also be found at www.asce.org.

The provisions of this document are written in permissive language and, as such, offer to the user a series of options or instructions but do not prescribe a specific course of action. Significant judgment is left to the user of this document.

This standard has been prepared in accordance with recognized engineering principles and should not be used without the user's competent knowledge for a given application. The publication of this standard by ASCE is not intended to warrant that the information contained therein is suitable for any general or specific use, and ASCE takes no position respecting the validity of patent rights. The user is advised that the determination of patent rights or risk of infringement is entirely his or her own responsibility.

A complete list of currently available standards is available in the ASCE Library (http://ascelibrary.org/page/books/s-standards).

ASCE STANDARD

ANSI/ASCE/EWRI

66-17

Management Practices for Control of Erosion and Sediment from Construction Activities

PUBLISHED BY THE AMERICAN SOCIETY OF CIVIL ENGINEERS

Library of Congress Cataloging-in-Publication Data

Names: American Society of Civil Engineers.
Title: Management practices for control of erosion and sediment from construction activities / American Society of Civil Engineers.
Description: Reston, Virginia : American Society of Civil Engineers, [2017] | "ANSI/ASCE/EWRI 66-17." | Includes bibliographical references and index.
Identifiers: LCCN 2017015047 | ISBN 9780784414224 (soft cover : alk. paper) | ISBN 9780784449407 (PDF)
Subjects: LCSH: Soil stabilization–Standards. | Building sites–United States–Standards. | Soil erosion–Prevention–Standards. | Sediment control–Standards. | Soil conservation–Standards.
Classification: LCC TA749 .A44 2017 | DDC 624.1/51363–dc23 LC record available at https://lccn.loc.gov/2017015047

Published by American Society of Civil Engineers
1801 Alexander Bell Drive
Reston, Virginia, 20191-4382
www.asce.org/bookstore | ascelibrary.org

This standard was developed by a consensus standards development process that has been accredited by the American National Standards Institute (ANSI). Accreditation by ANSI, a voluntary accreditation body representing public and private sector standards development organizations in the United States and abroad, signifies that the standards development process used by ASCE has met the ANSI requirements for openness, balance, consensus, and due process.

While ASCE's process is designed to promote standards that reflect a fair and reasoned consensus among all interested participants, while preserving the public health, safety, and welfare that is paramount to its mission, it has not made an independent assessment of and does not warrant the accuracy, completeness, suitability, or utility of any information, apparatus, product, or process discussed herein. ASCE does not intend, nor should anyone interpret, ASCE's standards to replace the sound judgment of a competent professional, having knowledge and experience in the appropriate field(s) of practice, nor to substitute for the standard of care required of such professionals in interpreting and applying the contents of this standard.

ASCE has no authority to enforce compliance with its standards and does not undertake to certify products for compliance or to render any professional services to any person or entity.

ASCE disclaims any and all liability for any personal injury, property damage, financial loss, or other damages of any nature whatsoever, including without limitation any direct, indirect, special, exemplary, or consequential damages, resulting from any person's use of, or reliance on, this standard. Any individual who relies on this standard assumes full responsibility for such use.

Errata: Errata, if any, can be found at http://dx.doi.org/10.1061/9780784414224.

ISBN 978-0-7844-1422-4 (soft cover)
ISBN 978-0-7844-4940-7 (PDF)
Manufactured in the United States of America.

24 23 22 21 20 19 18 17 1 2 3 4 5

CONTENTS

PREFACE

From a brief assessment of the major areas for the control of erosion and sediment (ESR), this standard is written as summarized as follows.

In assembling the first draft of this document, Michael Clar reviewed a number of public guidance documents, from which he abstracted liberally from the best of the available information.

For example, Chapter 2, Background, borrows liberally from the Delaware Erosion and Sediment Control Handbook.

Chapter 6, Erosion and Sediment Control Plan, incorporates a substantial amount of information from the International Erosion and Sediment Control Association (IECA 2003).

Chapter 7 was also largely adapted from another document, from the Delaware Department of Natural Resources and Environmental Control (DNREC, 2013).

Chapter 8, Construction Inspection and Maintenance, was adapted from documents used by New Castle County in Delaware and the Delaware Construction Inspection program.

Chapter 9, Enforcement, was adapted from documents developed by New Castle County, Delaware.

The units in these guidelines are Système International (SI) units followed by customary units.

Major Resource Documents

The developers of these guidelines provide a special acknowledgment of the following documents, which served as valuable resources for the development of the standard.

Delaware Sediment and Stormwater Program, Division of Soil and Water Conservation, Department of Natural Resources and Environmental Control (DNREC). (2013). *Delaware erosion and sediment control handbook*, Dover, DE.

Department of Health, Environmental Health Administration, Bureau of Environmental Quality, Watershed Protection Division. (2003). "District of Columbia standards and specifications for soil erosion and sediment control." Government Printing Office, Washington, DC.

ICEA (International Erosion Control Association). (2001). "Workbook: How to select, install and inspect construction site erosion and sediment control BMPs for NPDES storm water permit compliance." Steamboat Springs, CO.

Maryland Department of the Environment, Natural Resources Conservation Service, and Maryland Association of Soil Conservation Districts. (2011). "Maryland standards and specifications for soil erosion and sediment control." Water Management Administration, Baltimore.

Michigan Infrastructure Services, Design and Construction Division, Department of Management and Budget. (2002). "Michigan soil erosion and sedimentation control guidebook." Grand Rapids, MI.

New York State Department of Environmental Conservation Soil and Water Conservation Committee. (2005). "New York State standards and specifications for erosion and sediment control." New York Dept. of State, New York.

Nonpoint Source Management Program, Department of Environmental Protection. (2003). "The Florida stormwater, erosion, and sediment control inspector's manual." Florida Dept. of Environmental Protection, Tallahassee, FL.

Pennsylvania Department of Environmental Protection, Bureau of Waterways Engineering and Wetlands. (2012). "Erosion and sediment pollution control program manual." Harrisburg, PA.

Virginia Department of Environmental Quality (VDEQ), Division of Soil and Water Conservation. (1992). *Virginia erosion and sediment control handbook*, 3rd Ed., Richmond, VA.

ACKNOWLEDGMENTS

These guidelines were developed by the Management Practice for the Control of Erosion and Sediment (MPCES) Standards Committee of the Standards Development Council of the Environmental and Water Resources Institute (EWRI) of the American Society of Civil Engineers. The officers and members of the MPCES Standards Committee who balloted on the guidelines are listed as follows. The Development of the Design Guidebook (DDG) Subcommittee was involved with all revisions of the document. The Development of Regulatory Tools (DRT) Subcommittee was involved in the development of Appendix A.

MPCES Standards Committee

Donald W. Baker, P.E., M.ASCE, D.WRE, CPESC, *Past Chair of DDG*
Billy Barfield, Ph.D., P.E., M.ASCE
Michael T. Buechter, P.E., M.ASCE, *Member of DDG*
Michael L. Clar, P.E., M. ASCE, D.WRE, CFM, *Corresponding Editor, Former Chair, and Initial Member of DDG*
Shirley E. Clark, Ph.D., M.ASCE, D.WRE, *Past Secretary*
David D. Dee, Jr., P.E., M.ASCE, D.WRE, *Vice Chair and Member of DDG*
Darrell Eidson, P.E., M.ASCE, D.WRE, *Past Chair and Member of DDG*
Richard Field, P.E., M.ASCE, D.WRE, BCEE
Kendall F. Holbrook, *Chair of Performance-Based BMPs Subcommittee*
Karen Kabbes, P.E., *Member of DRT*
Conrad G. Keyes, Jr., Sc.D, P.E., P.S., Dist.M.ASCE, D.WRE, *Chair, DRT Past Chair, and eRoom Coordinator of DDG*
Eugene C. McCall, Jr., Ph.D, P.E., JD, M.ASCE, *Chair of DRT and Corresponding Editor for the final appendices*
Brian Noll, *Member of DDG*
Thomas P. O'Connor, P.E, M.ASCE, *Corresponding Editor, Section 5 Drawing Developer, and Chair of DDG*
Dennis L. Richards, P.E, D.WRE, F.ASCE, *Secretary, Former MPCES Representative to EWRI SDC, and Initial Member of DDG*
Timothy J. Ward, Ph.D, P.E., L.M. ASCE, F.ASCE, F.EWRI, M.NSPE
Stan E. Wildesen, P.E., M.ASCE, *Member of DRT*

CHAPTER 1
INTRODUCTION AND SCOPE

The successful implementation of an erosion and sediment control (ESC) program is a multifaceted undertaking that includes a mix of administrative, legal, and technical issues. This standard provides guidelines for personnel involved in the implementation of erosion and sediment control programs. These affected industries include land development and construction, engineering and environmental services, regulatory including plan reviewers, field inspectors, and administrative personnel.

The guidelines are organized into nine chapters that address the following topics:

- Importance of clean water (background)
- Regulations and permitting
- Erosion and sediment processes and site planning concepts
- Erosion and sediment control practices and standards
- Erosion and sediment control plan
- Erosion and sediment control guidance for specific development types
- Construction inspection and maintenance
- Enforcement.

Chapter 2, Background, summarizes the importance of clean water. It includes a brief summary of the environmental and economic impacts of sediment pollution. It also provides an overview of the environmental and economic advantages of compliance with erosion and sediment control regulations.

Chapter 3, Regulations and Permitting, summarizes the erosion and sediment control laws and regulations at the federal, state, and local levels. This chapter also provides a summary of the recommended permitting procedures.

Chapter 4, Control Processes and Site Planning Concepts, provides an overview of a number of erosion and sediment control processes and site planning concepts. The erosion processes that are described include raindrop, sheet, rill, gully, and channel erosion. Factors influencing erosion that are described include climate/rainfall, soils, topography, vegetative cover, and management/conservation practices. Site planning concepts that are described include planning the development to fit the site, limits of clearing and grading, natural drainage areas, area and duration of exposure, perimeter control practices, preventing excessive on-site damage, runoff velocities and runoff retention, stabilizing disturbed areas, and maintenance and follow-up programs.

Chapter 5, Control Concepts, Practices, and Standards, provides an overview of erosion and sediment control practices and standards. The available control practices are organized into the following functional groupings:

- Erosion control for graded areas
- Erosion control for slopes
- Erosion control for drainage ways
- Sediment barriers
- Construction entrance
- Sediment traps
- Sediment basin.

Chapter 6, Erosion and Sediment Control Plan, provides guidance related to the preparation of an erosion and sediment control plan. This chapter describes the four main elements of an erosion and sediment control plan: narrative, map/site plan, construction details, and engineering calculations. Guidelines are provided on both the preparation and evaluation of an erosion and sediment control plan.

Chapter 7, Control Guidance for Specific Types of Development, provides guidelines for specific types of construction activities that include

- Road construction projects
- Utility construction
- Commercial / industrial development
- Residential development.

Chapter 8, Construction Inspection and Maintenance, provides guidance related to inspection and maintenance of erosion and sediment control practices. This chapter provides a summary of the standard operating procedures for inspection of soil erosion and sediment control. It provides guidance directed specifically at erosion control inspectors and addresses conduct, compliance, and handling violations. Guidance is also provided for conducting site inspections. The topics described here are the inspection, tolerances, preparation for an inspection, preconstruction conference, office preparation, site inspection, and causes of noncompliance. Guidelines are also provided on inspection of individual erosion and sediment control practices.

Chapter 9, Enforcement, provides a summary of the standard operating procedure on the enforcement of soil erosion and sedimentation control.

The units used in these guidelines are SI units followed by customary units in parenthesis, except for the Model Ordinance in Appendix A (only in customary units).

1.1 APPROPRIATE DOCUMENTS

The ASCE standards used in Appendix E are listed herein and can be obtained from ASCE, 1801 Alexander Bell Drive, Reston, VA:

ASCE. "Regulated riparian model water code." ASCE/EWRI 40, Reston, VA.

ASCE. "Comprehensive transboundary water quality management agreement." ASCE/EWRI 33, Reston, VA.

ASTM standards referenced herein are as follows:

ASTM. "Standard test method for deterioration of geotextiles by exposure to light, moisture and heat in a xenon arc type apparatus." ASTM D4355, Developed by Subcommittee D35.02, Philadelphia, PA.

ASTM. "Standard test method for grab breaking load and elongation of geotextiles." ASTM D4632, Developed by Subcommittee D35.01, Philadelphia, PA.

Appropriate documents other than the standards cited here are in Appendix E.

CHAPTER 2
BACKGROUND

Natural erosion is generally considered to be due to the influence of climatic forces on the surface of the earth. Although one can learn from the processes of natural erosion, the practice of erosion control usually is limited to sites where human activities accelerate this natural process. The goal of erosion and sediment control regulations is not to stop natural erosion but only to control accelerated erosion caused by human activities so that there is no significant increase in sediment being discharged from a construction site over preconstruction conditions.

2.1 EROSION AND SEDIMENT PROBLEMS ASSOCIATED WITH CONSTRUCTION ACTIVITIES

The Delaware Sediment and Stormwater Program, Division of Watershed Stewardship, has published an excellent overview of this topic (DNREC 2013). The discussion that follows has been largely adapted from this document, with permission from the Delaware Department of Natural Resources and Environmental Control.

Erosion problems can be accelerated by a variety of human activities, including unrestricted development, overtaxed resources, removal of surface cover (such as vegetation), increased imperviousness (such as paving and rooftops) that increases runoff, and poor stewardship.

The principal effect construction activities have on the natural or geologic erosion process consists of exposing disturbed soils to precipitation, stormwater runoff, and the resultant shear stresses placed on the newly bare, unprotected soil. Grading and shaping of the land for construction or development purposes alters the soil cover and the soil in many ways, often detrimentally affecting on-site drainage and storm runoff patterns and eventually the off-site stream and streamflow characteristics. Protective vegetation is reduced or removed, topography is altered, and the removed soil is stockpiled, often without protective cover. In effect, the physical properties of the soil are changed.

Construction-related erosion and sedimentation can cause problems for down-slope property owners, create nuisance problems on adjacent streets, clog streams and storm drains, result in turbidity plumes in downstream water bodies, and cover sensitive habitat areas with sediment. The cumulative toll on the environment can be devastating. Uncontrolled erosion is costly; violates state and federal pollution laws; endangers downstream aquatic ecosystems; exposes developers, contractors, and landowners to legal liabilities; and provides evidence to those who argue that the development process is out of control. Surface water pollution, aquatic ecosystem damage, channel and reservoir siltation, and damage to public facilities, as well as to private property, are some of the many examples of problems caused by uncontrolled erosion and sedimentation.

Some of the potential hazards associated with construction activities include the following:

1. A large increase in the amount of bare or poorly protected soils exposed to storm runoff and subject to soil erosion
2. Increased volumes of storm runoff, accelerated erosion and sediment yield, and higher peak flows caused by
   - Removal of protective vegetation
   - Exposure of underlying soil or geologic formations that are less pervious and/or more erodible than the original surface
   - Reduced capacity of exposed soils to absorb rainfall due to compaction caused by heavy equipment
   - Enlarged drainage areas caused by grading operations, diversions, and street construction
   - Prolonged exposure of unprotected disturbed areas and stockpiles due to scheduling problems and/or delayed construction
   - Shortened times of concentration of surface runoff caused by altering steepness, distance, and surface roughness, and through installation of "improved" storm drainage facilities
   - Increased impervious surfaces associated with the construction of streets, buildings, sidewalks and paved driveways, and parking lots.
3. Alteration of the groundwater and infiltration regime through soil compaction that may adversely affect drainage systems, slope stability, and survival of existing and/or newly established vegetation
4. Exposure of subsurface materials (bedrock, hydric, or clay soils, compacted or low-permeability soils) that are rocky, acid, droughty, or otherwise unfavorable to the establishment of vegetation
5. Adverse alteration of watershed boundaries and surface runoff patterns by construction and development.

Although streams and rivers naturally carry sediment loads, erosion from construction sites and runoff from developed areas can elevate these loads to levels well above those in undisturbed watersheds. It is generally acknowledged that erosion rates from construction sites are much greater than from almost any other land use. Results from both field studies and erosion models indicate that erosion rates from construction sites are typically about an order of magnitude larger than row crops and several orders of magnitude greater than rates from well-vegetated areas such as forests or pastures (Yorke and Herb 1976, Dillaha et al. 1989, Meyer et al. 1971). Wolman and Schick (1967) studied the impacts of development on fluvial systems in Maryland and found that sediment yields in areas undergoing construction were 1.5 to 75 times greater than detected in natural or agricultural catchments. They also summarized the potential impacts of construction on sediment yields by stating that the equivalent of many decades of natural or even agricultural erosion may take place during a single year from areas cleared for construction.

Similar impacts from stormwater runoff have been reported in a number of other studies. For example, Daniel et al. (1979) monitored three residential construction sites in southeastern Wisconsin and determined that annual sediment yields were more than 19 times the yields from agricultural areas. In northern Idaho, the erosion rate per unit area of road surface averaged 220 times the erosion rate of undisturbed areas over a six-year period (Megahan and Kidd 1972).

Other studies have documented increased surface erosion following road construction but at increases smaller than the 220-fold increase reported in the 1972 study (Megahan 1986). A highway construction project in West Virginia disturbed only 4.2% of a 12.30 km^2 (4.75 sq mi) basin, but this resulted in a three-fold increase in suspended sediment yields (Downs and Appel 1986). During the largest storm event it was estimated that 80% of the sediment in the stream originated from the construction site. Another study evaluated the effect of 117 ha (290 ac) of highway construction on watersheds ranging in size from 13 to 98 km^2 (5 to 38 sq mi). Suspended sediment loads in the smallest watershed increased by 250%, and the estimated sediment yield from the construction area was 14,000 kg/ha (37 tons/ac) over a two-year period (Hainly 1980).

Yorke and Herb (1978) evaluated nine sub-basins in the Maryland portion of the Anacostia watershed. Average annual suspended sediment yields for construction sites ranged from 2,600 to 37,000 kg/ha (7 to 100 tons/ac). Daniel et al. (1979) identified flow, followed by flow rate, as the most influential factor controlling the sediment loadings from residential construction sites. In evaluating subbasins at a scale of between 0.91 and 57 km^2 (0.35 and 22 sq mi), Yorke and Herb (1976) determined that sediment yields from developed "urban land" were on the upper end of the range for cultivated land (1,400 kg/ha; 3.7 tons/ac) as compared to 240 to 1,600 kg/ha (0.65 to 4.3 tons/ac). However, the source of most of the sediment in the urban land streams was identified as stream-channel erosion, not the developed properties themselves. Later studies, such as Nelson (1996), as cited in Pitt et al. (2007), demonstrate the importance of rainfall intensity on suspended solids generation and median particle size in construction site runoff (rain intensity < 6.4 mm/h [0.25 in/h], suspended solids = 400 mg/L and median size = 3.5 um; rain intensity > 25 mm/h [1 in/h], suspended solids = 25,000 mg/L and median size = 8.5 um).

Other studies have shown that stream reaches affected by construction activities often extend well downstream of the construction site. For example, between 4.8 and 5.6 km (3.0 and 3.5 mi) of stream below construction sites in the Patuxent River watershed were observed to be impacted by sediment inputs (Fox 1974a,b as cited in Klein 1979). Many studies suggest that the most effective way to reduce sediment losses is to prevent erosion.

The environmental and economic consequences of allowing such activities to go unchecked are discussed in the following sections.

2.2 ENVIRONMENTAL IMPACTS OF SEDIMENT

Stormwater discharges and the associated sediment loads generated while construction activities are occurring have a potential for serious water quality impacts. The biological, chemical, and physical integrity of the waters may become severely compromised. For example, a number of pollutants are preferentially absorbed into mineral or organic particles found in fine sediment. The erosion and delivery of sediment into aquatic ecosystems is the primary pathway for delivering key pollutants, such as nutrients (particularly phosphorous), metals, and organic compounds (e.g., Novotny and Chesters 1989). It has been estimated that 80% of the phosphorous and 73% of the Kjeldahl nitrogen in streams may be associated with eroded sediment (Fennessey and Jarrett 1994). Even more startling is the apparent ability of sediment to act as long-term memory or storage media for toxicants. Studies show that pollutants such as DDT, DDE, PCBs, and chlordane, whose use has been banned or highly restricted, can still be found at detectable levels in sediment deposited years ago in the bottom of streams and rivers.

Water quality parameters that reflect the level of sediment yield are turbidity and suspended solids. As turbidity increases within a stream environment, photosynthetic activity may decrease, with a subsequent potential decrease in available free oxygen necessary to support aquatic life. An increase in the concentration of suspended solids may destroy water supplies for human and wildlife consumption, as well as degrading feeding and nesting habitats. Implementation of erosion control features consistent with sound construction operations can minimize the adverse impacts associated with increased sediment yield.

2.2.1 In-stream and Off-stream Impacts. Some of the in-stream and off-stream impacts of turbidity and sedimentation are identified as follows:

Sedimentation Impacts – In-stream Damages

- Destroys spawning areas, food sources, habitat
- Directs toxicity to wildlife
- Degrades lakes
- Silts navigation channels
- Impacts commercial fisheries
- Reduces water storage capacities.

Sedimentation Impacts – Off-stream Damages

- Increases flood hazards
- Increases water treatment costs
- Decreases capacity in conveyance facilities.

Turbidity from Suspended Solids

- Transports nutrients, pesticides, bacteria, toxic substances
- Harms aquatic wildlife
- Reduces beneficial uses.

2.2.2 Specific Environmental Impacts. These in-stream and off-stream impacts may be translated into specific environmental impacts, as summarized as follows:

- *Resource Base Impacts* — Loss of soil as a resource results in the elimination of potential for future use and decreased biological diversity.
- *Agricultural Impacts* — Loss of soil results in reduced crop production and higher management costs (such as seed and fertilizer).
- *Water Quality Impacts* — Sediment can cause damage to fish and wildlife resources, water supply quality, recreational values, and habitat values.
- *Air Quality Impacts* — Fugitive dust can cause public health and safety problems.

Where construction activities are intensive, the localized impacts of water quality may be severe because of high pollutant loads, primarily sediments. Siltation is the primary cause of impaired water quality in rivers and the second-largest cause of impaired water quality in lakes (USEPA 1992). Excess sediments are associated with increased turbidity and reduced light penetration in the water column, as well as more long-term effects associated with habitat destruction and increased difficulty in filtering drinking water. Numerous studies have shown that fine sediment (fine sand or smaller) can affect aquatic ecosystems adversely by reducing light penetration, impeding sight-feeding, smothering benthic organisms, abrading gills and other sensitive structures, reducing habitat by clogging interstitial spaces within a stream bed, and reducing the inter-gravel dissolved oxygen by reducing the permeability of the bed material (e.g., Everest et al. 1987, Meehan 1991). In an aquatic environment, the general effect of fine-graded sediments, such as clays, silts, and fine sands, is to reduce drastically both the kinds and the numbers of organisms present.

Coarser-grained materials also blanket bottom areas and suppress aquatic life found in these areas. Where currents are sufficiently strong to move the bedload, the abrasive action of these materials in motion accelerates channel scour and has an even more severely deleterious effect on aquatic life. Introduction of coarse sediment (coarse sand or larger) or large amounts of fine sediment is also a concern because of the potential of filling lakes and reservoirs as well as clogging stream channels (e.g., Paterson et al. 1993). Large inputs of coarse sediment into stream channels initially will reduce stream depth and minimize habitat complexity by filling in pools (MacDonald et al. 1991).

For most construction sites the primary concern is fine sediment, because the primary erosion and transport processes are rain-splash, rills, typically less than 0.3 m (1 ft) deep, and sheet wash (Haan et al. 1994). Construction sites in steep areas or along a stream bank, however, may initiate landslides, debris flows, or other types of mass-wasting events (e.g., Megahan 1983). In these cases coarse sediment inputs may be of greatest concern. Construction sites also can generate other pollutants associated with wastes on site, such as sanitary wastes, concrete truck washout, or chemical spills.

Evidence is now beginning to mount that the ability of a stream to maintain its health is related to the impervious areas within its watershed. Urbanization of a watershed increases the impervious surfaces and increases the pollutant load. Numerous studies indicate that stream impairment becomes evident once a watershed reaches an impervious cover threshold of 10% (Schueler 1995).

2.3 ECONOMIC IMPACTS OF SEDIMENT

More than 3.6×10^{12} kg (4 billion tons) of sediment is estimated to reach the ponds, rivers, and lakes of the United States each year, and approximately 9.1×10^{11} kg or 0.9×10^{12} billion tons of this sediment is actually carried all the way to the ocean. Although these latter quantities may appear to be small compared to the total, they could represent more than one-half of the sediment load carried by many streams draining small subwatersheds that are undergoing development.

Excessive quantities of sediment cause costly damage to waters and to private and public lands. Obstruction of stream channels and navigable rivers by masses of deposited sediment reduces their hydraulic capacity, which, in turn, causes an increase in subsequent flood crests and a consequent increase in the frequency of damaging storm events. The aesthetic attraction of many streams, lakes, and reservoirs used for swimming, boating, fishing, and other water-related recreational activities has been seriously impaired or destroyed by bank cutting and channel scour, accelerated by these higher flood stages induced by sedimentation.

Sediment fills drainage channels, especially along highways and railroads, and plugs culverts and storm drainage systems, thus necessitating frequent and costly maintenance. Municipal and industrial water supply reservoirs lose storage capacity, the usefulness of recreational impoundments is impaired or destroyed, navigable channels must be dredged continually, and the cost of filtering muddy water preparatory to domestic or industrial use becomes excessive—and sometimes exorbitant. The added expense of water purification in the United States, because of sedimentation, amounts to millions of dollars each year.

Every year, the U.S. Army Corps of Engineers and private operators dredge millions of cubic meters (y^3) of material to create and maintain navigable highways and harbors. Without such activity, the natural processes of sediment deposition and shoaling would make many waterways and port facilities impassable by most large commercial and defense vessels. The process of keeping these waters passable is an expensive one. In addition to the financial cost of dredging, the disposal of dredged material has become more difficult with the passage of laws to protect the marine environment and the consequent dwindling supply of suitable places to put it. Thus, consideration of alternatives to dredging has become an issue with serious implications for the United States' maritime trade and naval installations.

In the past, the damage caused by erosion and sedimentation has been quantified in terms of dollars spent to dredge navigational channels, loss of reservoir capacity, and so on. More recently, efforts have concentrated on the qualitative cost. Responsible development requires that steps be taken to control erosion and sedimentation from construction sites in order to minimize the environmental and economic impacts associated with such activities. For example, it could be constructive to have "homebuilders to pay penalties for stormwater violations" and "the settlements [to] require companies to develop improved pollution prevention plans ... [and] submit annual reports to the USEPA" (Northcut 2008).

2.4 OTHER POTENTIAL NONSEDIMENT POLLUTANTS

Research has shown that there are also potential nonsediment pollutants likely to be present in stormwater in significant quantities. Potential pollutants that are commonly associated with construction activities include the following.

2.4.1 Metals. Galvanized metal, painted surfaces, and pressure-treated lumber comprise many of the surfaces exposed to stormwater as a result of construction activity. These coatings and treatments contain metals that enter stormwater as the surfaces corrode, flake, dissolve, decay, or leach. Acid rain can accelerate these processes.

2.4.2 Nutrients. Nutrients, especially nitrogen and phosphorus, run off from construction sites as a result of fertilizers applied to encourage growth of temporary or permanent vegetation. In addition, nutrients leach from the decaying of mulches and straw-bale barriers.

2.4.3 Pesticides. Herbicides, insecticides, and rodenticides are commonly used at construction sites. The unnecessary or improper application of these pesticides may result in direct contamination, indirect pollution through drift, or the transport of soil surfaces into water.

2.4.4 Spills and Illegal Dumping of Construction Materials. Petroleum products, pesticides, and other synthetic organic compounds (glues, sealants, solvents, etc.) are used widely at construction sites and may be improperly stored and disposed. Deliberate dumping of these materials, which can migrate into surface or ground water resources, is a direct violation of the Clean Water Act (see Chapter 3). On parking lot or highway construction projects, the application of diesel fuel to the contact surfaces of the "hot mix asphalt" application and transport vehicles is a common practice that should be discontinued.

2.4.5 Miscellaneous Wastes. Miscellaneous wastes include wash from concrete mixers; solid wastes resulting from the clearing and grubbing of vegetation; wood and paper materials derived from packaging of building products; food containers such as paper, aluminum, and steel beverage cans; and sanitary wastes. In addition to erosion and sediment controls, a management plan must address the other potential pollutant sources that may exist on a construction site. These controls include proper disposal of construction site waste, compliance with applicable state or local waste disposal, sanitary sewer or septic system regulations, control of off-site vehicle tracking, and control of allowable nonstormwater discharges.

2.5 ECONOMIC ADVANTAGES OF COMPLIANCE

Compliance with applicable federal, state, and local erosion control and stormwater pollution prevention regulations can result in economic advantages, particularly if one considers the costs of noncompliance in terms of stop work orders, fines, and lawsuits.

The advantages of compliance at construction site may include the following:

- Potential for fines for noncompliance can be reduced or eliminated.
- Stabilized slopes require less repair and are safer for maintenance crews.
- Reducing short- and long-term erosion will result in less soil loss.
- Negative public opinion, which can result in enforcement actions, can be minimized.
- Liability exposure can be decreased.

Some measurable costs resulting from noncompliance include

- Removal of silt deposits from storm drains and ultimately, water bodies
- Reduction of water storage capacity of reservoirs
- Increased water treatment costs
- Increase in flooding hazards.

2.6 ENVIRONMENTAL ADVANTAGES OF COMPLIANCE

When groups attempt to address one aspect of environmental quality positively, there are usually many side benefits. Erosion and sediment control are no exceptions. Some of the environmental benefits of effective erosion and sediment control can be

- Protection of fish spawning areas, their food sources and habitat
- Reduction of toxic materials that are introduced into the environment by their attachment to and transport by sediment particles
- Lowered impact on commercial fisheries
- Improved water quality and storage capacities in lakes and reservoirs
- Protection of soil as a resource, thereby maintaining a future of diverse and beneficial uses
- Protection of human and wildlife uses of receiving waters.

CHAPTER 3
REGULATIONS AND PERMITTING

Erosion and sediment control requirements are mandated and regulated by a number of federal, state, and local laws and regulations. This chapter provides a brief summary and discussion of the nature and extent of these regulatory requirements. Appendix B and C have provided the appropriate federal and state regulations, respectively.

The generally stated goal and objectives of federal, state, and local regulations are to safeguard persons, protect property, and prevent damage to the environment resulting from uncontrolled construction operations. The regulations are also generally intended to promote the public welfare by guiding, regulating, and controlling the design, construction, use, and maintenance of any development or other activity that disturbs or breaks the topsoil or results in the movement of earth on land.

3.1 MODEL ORDINANCE

Appendix A provides a suggested erosion and sediment model ordinance, which has been modified by this committee based on a model ordinance developed by the U.S. Environmental Protection Agency Office of Wetlands, Oceans, and Watersheds (USEPA 2015).

The principal components of this model ordinance include

- Declaration of purpose and objectives
- Definitions
- Application guidelines
- Permit requirements and approval process
- Contents of an erosion and sediment control plan
- Design requirements
- Inspection and monitoring
- Maintenance requirements
- Enforcement procedures
- Training and certification.

3.2 FEDERAL REGULATIONS

Although a number of state and local erosion and sediment control regulations were adopted and implemented prior to the adoption of the Clean Water Act, most state and local regulations have been adopted pursuant to the provisions of this act and some related regulations that include

- U.S. EPA Clean Water Act (Sections 401, 402, 404)
- National Pollutant Discharge Elimination System (NPDES) Program Phases I and II
- Pollution Prevention Act
- Coastal Zone Re-authorization Act
- Endangered Species Act.

Appendix B provides a detailed discussion of these federal regulations. This section only focuses on the NPDES program provisions of the Clean Water Act.

3.2.1 NPDES Program (Phases I and II). In 1990 EPA published its final "Rule for regulation of stormwater discharges." This rule (40 CFR 121.26), known as the Phase I Rule, established requirements for the permitting and management of stormwater discharges from large (serving a population of 250,000 or more) and medium-sized (serving a population of 100,000 to 250,000) municipalities, numerous industries (based on the Standard Industry Classification code), and construction sites disturbing 2 ha (5 ac) or more. The permits are administered by the EPA (or certain delegated state agencies) under the NPDES program.

Under the Phase I Rule, coverage under an EPA construction general permit (CGP) must be sought by the operator of a construction activity that

- Will disturb 2 ha (5 ac) or greater, or
- Will disturb less than 2 ha (5 ac) but is part of a larger common plan of development or scale whose total land-disturbing activities total 2 ha (5 ac) or greater (or is designated by the NPDES permitting authority), and
- Will discharge stormwater runoff from the construction site into a municipal separate storm sewer system (MS4) or waters of the United States.

In December 1999, EPA finalized the Phase II Rule regulations, which require controls on stormwater discharges from a broader sector of municipalities, industries, and construction sites. Specifically for construction, the Phase II Rule requires construction sites disturbing equal to or greater than 0.40 ha (1 ac) and less than 2 ha (5 ac) to control pollutants in stormwater runoff. Construction activity disturbing less than 0.40 ha (1 ac) requires a permit if it is part of a larger common plan of development or sale disturbing a total of 0.40 ha (1 ac) or greater or is individually designated for permit coverage by the NPDES permitting authority.

The Phase II Rule requires, nationally, operators of Phase II construction sites to obtain an NPDES permit and implement best management practices (BMPs) to minimize pollutant runoff. For the Phase II construction program, EPA has taken an approach similar to the Phase I approach where the program requirements are not fully defined in the rule but rather in the NPDES stormwater permit issued by the NPDES permitting authority. The Phase II requirements are similar to the following three main requirements of EPA's CGP for Phase I:

- Submittal of a notice of intent (NOI) that includes general operator and site information
- Development and implementation of a stormwater pollution prevention plan (SWPPP) with appropriate BMPs to minimize the discharge of pollutants from the site prior to initiating any on-site activities, and

- Submittal of a notice of termination (NOT) when final stabilization of the site has been achieved as defined in the permit, stormwater runoff is no longer being discharged, or another operator has assumed control of the site.

A copy of EPA's overview of the Storm Water Phase II Final Rule for construction sites can be obtained from the EPA website: http://www.epa.gov/sites/production/files/documents/fact3-0.pdf.

The SWPPP is an evolving document. It has been suggested that there are five stages of a successful plan:

Stage 1: Initial preparation
Stage 2: Contractor review and modifications
Stage 3: Preconstruction perimeter controls installation
Stage 4: Modifications and updates during construction
Stage 5: Filing the NOT and storing the SWPPP.

3.2.2 Phase II Municipal Requirements. Phase II municipalities must apply for a NPDES stormwater permit, develop a Storm Water Management Plan (SWMP) that includes six minimum control measures, implement the SWMP using appropriate BMPs, develop measurable program goals, and evaluate program effectiveness. The applicable standards for Phase II municipalities are to reduce the discharge of pollutants to the "maximum extent practicable" (MEP), protect water quality, and satisfy the water quality requirements of the Clean Water Act.

The six minimum control measures that have been documented by EPA for an NPDES stormwater permit are

1. Public education and outreach
2. Public participation/involvement
3. Illicit discharge detection and elimination
4. Construction site runoff control
5. Postconstruction runoff control
6. Pollutant prevention/good housekeeping.

3.3 STATE AND LOCAL REGULATIONS

State and local regulations related to erosion and sediment (E&S) control generally include the 10 principal components of the model ordinance (see Appendix A) and are briefly described following. They are subject to some variance from state to state. Appendix C identifies the current erosion and sediment control regulatory agency for each state.

3.3.1 Declaration of Purpose and Objectives. This component of the regulations generally describes the basis and need for the regulations and the goals that the regulations seek to accomplish. This aspect has been described in detail in Chapter 2.

3.3.2 Definitions. This element of the regulations is self-explanatory but becomes essential to the implementation of the regulations.

3.3.3 Application Guidelines. This element of the regulations describes what activities are subject to the regulations and also identifies any activities that may be exempt from the regulations. This element also may identify any waivers or variances that apply as well as the process for applying for and obtaining a variance.

3.3.4 Permit Requirements and Approval Process. This element is very important, because it provides the required guidance to the regulated community on the permitting requirements and the approval process.

3.3.5 Contents of an Erosion and Sediment Control Plan. The contents of an erosion and sediment control plan are described in detail in Chapter 6 of this standard. The major components of an E&S plan are identified and include

- Narrative
- Map / site plan
- Construction details
- Supporting calculations.

3.3.6 Design Requirements. The design requirements usually are provided in the regulations as specific design criteria and/or minimum standards and specifications. The design criteria may include the design storm for sizing the facility and the spillway works, and allowable velocities, if applicable.

3.3.7 Inspection and Monitoring. The regulations also may address any required inspection or monitoring requirements. Chapter 8 provides detailed inspection and maintenance procedures for ESC practices.

3.3.8 Maintenance Requirements. Regulations often specify the required maintenance schedules for the structural ESC practices. Guidance on ESC practice maintenance is provided in Chapter 8.

3.3.9 Enforcement Procedures. Enforcement procedures often are described in ESC regulations. Chapter 9 of this standard describes the typical enforcement procedures that are in use today.

3.3.10 Training and Certification. Training and certification requirements are not always included in local regulatory programs. However, there is increasing recognition of the relatively specialized nature of ESC technology. One of the objectives of this standard is to help provide some basic materials to address this need in the regulated community.

CHAPTER 4
CONTROL PROCESSES AND SITE PLANNING CONCEPTS

This chapter provides an overview of a number of erosion and sediment processes and site planning concepts. The erosion processes that are described include raindrop erosion, sheet erosion, rill erosion, gully erosion, and channel erosion. Factors influencing erosion that are described include climate / rainfall, soils and soil erodibility, topography, vegetative cover, and management / conservation practices. Site planning principles that are described include planning the development to fit the site, limits of clearing and grading, natural drainage areas, area and duration of exposure, perimeter control practices, preventing excessive on-site damage, runoff velocities and runoff retention, stabilizing disturbed areas, and maintenance and follow-up programs.

4.1 EROSION PROCESSES

Soil erosion is the process by which the land's surface is worn away by the action of wind, water, ice, and/or gravity. Natural or geologic erosion has been occurring at a relatively slow rate since the earth was formed and is a tremendous factor in shaping the earth as we know it today. Except for some cases of shoreline and stream-channel erosion, natural erosion occurs at a very slow and uniform rate and remains a vital factor in maintaining environmental balance.

Water-generated erosion is unquestionably the most severe type of erosion, particularly in developing areas; it is, therefore, the problem that this guideline primarily addresses. It is helpful to think of the erosive action of water as the effects of the energy developed by rain as it falls or as the energy derived from its motion as it runs off the land surface, which these guidelines primarily address. The force of falling raindrops is applied vertically, and force of flowing water is applied horizontally. Although the direction of the forces created is different, they both perform work in detaching and moving soil particles.

Water-generated erosion can be broken down into the following five stages: (1) raindrop erosion, (2) sheet erosion, (3) rill erosion, (4) gully erosion, and (5) channel erosion. Fig. 4-1 illustrates the five stages of erosion.

4.1.1 Raindrop Erosion. Raindrop erosion is the first effect of a rainstorm on the soil. Raindrop impact dislodges soil particles and splashes them into the air (Fig. 4-2). These detached particles are then vulnerable to the next type of erosion.

4.1.2 Sheet Erosion. Sheet erosion is the erosion caused by the shallow flow of water as it runs off the land. These very shallow moving sheets of water are seldom the detaching agent, but the flow transports soil particles that are detached by raindrop impact and splash. The shallow surface flow rarely moves as a uniform sheet for more than a few meters or feet on land surfaces before concentrating in the surface irregularities.

4.1.3 Rill Erosion. Rill erosion is the erosion that develops as the shallow surface flow begins to concentrate in the low spots of the irregular contours of the surface. As the flow changes from the shallow sheet flow to deeper flow in these low areas, the velocity and turbulence of flow increase. The energy of this concentrated flow is able to both detach and transport soil materials. This action begins to cut small channels of its own. Rills are small but well-defined channels that are, at most, only a few inches or centimeters in depth. They are easily removed by harrowing or other surface treatments.

4.1.4 Gully Erosion. Gully erosion occurs as the flow in rills comes together in larger and larger channels. The major difference between gully and rill erosion is a matter of magnitude. Gully erosion generally is differentiated from rill erosion in that gullies have eroded to the point where they cannot be smoothed by normal tillage operations. Gullies are wider and can transport rock and more readily erode down to bedrock. When gullies are present, large equipment and special techniques are required for stabilization.

4.1.5 Channel Erosion. Channel erosion occurs as the volume, velocity, and duration of flow causes movement of the stream bed and bank materials.

4.2 FACTORS INFLUENCING EROSION

The erosion potential of any area is determined by five principal factors (climate/rainfall, soils and soil erodibility, topography—specifically flow path length and slope—vegetative cover, and conservation practices).

Although each of these factors is discussed separately in Sections 4.2.1 through 4.2.5, they are interrelated in determining erosion potential. The five factors are the primary factors that influence erosion and have been expressed mathematically in the Universal Soil Loss Equation (USLE) (USDA 1978), which is a semi-empirical equation based on 10,000 plot-years of data.

$$A = R \times K \times LS \times C \times P \tag{4-1}$$

where

A = long-term average annual soil loss in metric tons/ha (tons/ac)
R = rainfall and runoff factor
K = soil erodibility factor
LS = slope-length gradient factor
C = crop/vegetation and management factor and
P = support practice factor.

This historic equation is presented to show the different variables involved. For estimating soil loss and erosion, the latest version of the Revised Universal Soil Loss Equation 2

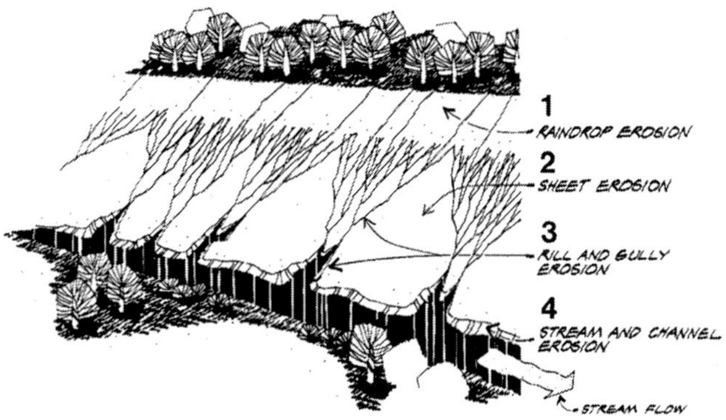

FIGURE 4-1. Basic Erosion Processes
***Source*: AISWCD (2013); reproduced with permission from the Association of Illinois Soil and Water Conservation Districts.**

FIGURE 4-2. Raindrop Erosion
***Source*: USDA Natural Resources Conservation Service.**

(RUSLE 2) is available from the U.S. Department of Agriculture (2016). RUSLE 2 applies an integration process, using daily factor values, and yields more meaningful results.

4.2.1 Climate/Rainfall (R). The frequency, intensity, and duration of rainfall are fundamental factors in determining the amount of runoff produced in a given area. As both the volume and velocity of runoff increase, the capacity of runoff to detach and transport soil particles also increases. Where storms are frequent, intense, or of long duration, erosion risks are higher. Seasonal changes in temperature, as well as variations in rainfall, help to define the high erosion risk period of the year. When precipitation falls as snow, no immediate erosion should take place. However, when the temperature rises, melting snow adds to runoff, and erosion hazards are higher. Rain falling on previously deposited snow can result in large rates of runoff, as the snow melted by the rainfall contributes to the runoff. Because the ground is still partially frozen, its absorptive capacity is reduced. Frozen soils are relatively erosion-resistant. However, soils with high moisture content are subject to uplift by freezing action and are usually very easily eroded upon thawing, because they are not as dense.

Climate influences the types of vegetation native to the area and consequentially the relative ease or difficulty of establishing both temporary and permanent vegetative cover. The average annual temperatures also would determine the earliest and latest dates for seeding and establishing vegetative cover.

R values, the measure of rainfall energy, are incorporated into RUSLE 2 but also can be read from maps supplied by the U.S. EPA in the Stormwater Phase II Rule: Construction Rainfall Erosivity Waiver (USEPA 2001). A review by Clark et al. (2009) of the R value calculation methods showed differences between the values read from the USDA isoerodent maps and those calculated using the equation supplied by the USEPA. The Clark et al. study also highlighted the impacts of the calculation methodology on the sizing and maintenance interval for sedimentation ponds.

4.2.2 Soils and Soil Erodibility (K). Soil characteristics that influence the potential for erosion by rainfall and runoff are those properties that affect the infiltration capacity of a soil and those that affect the resistance of the soil to detachment and being carried away by raindrop splash or flowing water.

Soil erodibility is the propensity for soil particles to become detached by actions of water or wind. The K factor is a function of soil texture, organic matter content, soil structure, and permeability. In addition, it is expressed as numerical values in the USDA/NRCS (2001b) handbook.

Soils containing high percentages of fine sands and silt are typically the most erodible. As the clay and organic matter content of these soils increases, the erodibility decreases. Clays act as a cohesive binder to soil particles, thus reducing erodibility. However, although clays have a tendency to resist erosion, once eroded, they are easily transported by water. Soils high in organic matter have a more stable structure, which improves their permeability and infiltration capacity. Such soils resist raindrop detachment and infiltrate more rainwater. Well-drained and well-graded gravel and gravel-sand mixtures are usually the least erodible soils. Soils with high infiltration rates and permeability either prevent or delay and reduce the amount of runoff.

4.2.3 Topography (LS). The size, shape, and slope characteristics of the land surface influence the amount and rate of runoff. As both slope length and gradient increase, the velocity of runoff increases and the potential for erosion is magnified as the resultant shear stresses on the soil increase. Slope orientation also can be a factor in determining erosion potential. For example, a slope that faces south and contains drier soils may have such poor growing conditions that vegetative cover is difficult to reestablish.

The degree to which the flowpath length and slope (LS factor) play a role in erosion can be calculated using USDA/NRCS charts (USDA/NRCS 2001b). Flowpath length is the distance along a flowpath to a point where deposition is first likely to occur. Slope steepness is the ratio of horizontal distance to vertical rise (e.g., 3:1 slope); percentage (e.g., 33% slope); or degrees (18° slope).

In general

- The effect of flowpath length is not as great as effect of slope steepness.
- Long slopes and especially long steep slopes should not be constructed.
- Steep slopes should be broken up by creating contour diversions or benches.
- A convex slope shape increases runoff and magnifies slope erosion.
- A concave slope shape enhances infiltration and reduces erosion.

Aspect or orientation of slope can be important with respect to vegetation establishment and moisture content.

4.2.4 Vegetative Cover (C). Vegetative cover plays an extremely important role in controlling erosion as it provides the following five benefits:

1. Shields the soil surface from raindrop impact
2. Holds soil particles in place by root system
3. Maintains the soil's capacity to absorb water
4. Slows the velocity of runoff
5. Removes subsurface water between rainfalls through the process of evapotranspiration.

The rate of erosion is related to the amount of permanent or vegetative cover (C). The functions of the cover are to

- Reduce rainfall impact on soil
- Reduce surface water velocities
- Enhance infiltration
- Filter sediment in surface runoff
- Retain soil particles in place and reinforce soil structure
- Promote permanent vegetation establishment.

By limiting and staging the removal of existing vegetation and by decreasing the area and duration of exposure, soil erosion can be reduced significantly. Special consideration should be given to the maintenance of existing vegetative cover on areas of high erosion potential, such as moderately to highly erodible soils, steep slopes, drainage ways, and the banks of streams.

4.2.5 Conservation Practices (P). Conservation practices should be controllable, imaginative, experience-driven, and interactive. One such factor is slope roughening perpendicular to the direction of runoff. For the most part, this factor can (USDA 1978)

- Enhance the factors of cover and soil texture
- Mitigate the influence of rainfall and runoff
- Modify velocity flow path length and steepness.

4.3 SEDIMENTATION

During a typical rainfall-runoff event, runoff builds up rapidly to a peak and then diminishes. Soil particles are eroded and transported principally during these higher-flow periods. During lower flows, the velocity of runoff decreases and the transported materials are deposited and may be transported again in later peak flows. In this way, eroded soil can be carried downslope or downstream intermittently and progressively from its source or point of origin trough to several temporary storage locations on the land surface or in a channel. This process is known as sedimentation.

It should be recognized that, ideally, sedimentation is a natural process of dynamic equilibrium. A certain amount of sedimentation and sediment transport occurs in even the most stable streams and rivers. However, when human activities change the sediment loads and/or the hydrology of a watershed, streams can quickly go out of equilibrium. Although the stream may eventually adjust to these changes, more sediment might be carried away from the banks and bed in the process, only to be deposited in downstream receiving waters. The potential environmental and economic impacts of sediment are discussed in Chapter 2, sections 2.5 and 2.6.

4.4 SITE PLANNING CONCEPTS

For an erosion and sediment control program to be effective, provisions for control measures should be made in the planning stage. These planned measures, when correctly implemented during construction, should result in erosion and sediment control, minimizing environmental degradation. The following principles should be used to the maximum extent possible (MDE 1994a–1994r):

- Plan the development to fit the site.
- Determine the limits of clearing and grading.
- Divide the site into natural drainage areas.
- Minimize the area and duration of soil exposure.
- Apply perimeter control practices.
- Apply erosion control practices to prevent excessive on-site damage.
- Keep runoff velocities low and retain runoff on site.
- Stabilize disturbed areas immediately.
- Implement a thorough maintenance and follow-up program.

4.4.1 Plan the Development to Fit the Site. The first planning concept of erosion and sediment control is to plan the development to fit the particular soils, drainage patterns, and natural vegetation of the site. Detailed planning should be employed to ensure that roadways, buildings, and other permanent features of the development conform to the natural characteristics of the site. Large graded areas should be located on the most level portion of the site. Slope length and gradient are key elements in determining the volume and velocity of runoff, the shear stress applied to the soil, and its associated erosion. As both slope length and steepness increase, the rate of runoff increases, and the potential for erosion is magnified. Steep vegetated slopes should be left undisturbed. Areas with slope and soil limitations should not be used unless strict conservation practices are adhered to. For instance, where it is necessary to build on long steep slopes, the practices of benching, terracing, or constructing diversions should be used. Areas subject to flooding should be avoided or used as part of the stormwater management system. Floodplains should be kept free from filling and construction activities, as they temporarily store excess runoff, thus helping to avoid erosion and flooding problems downstream. Erosion control, development, and maintenance costs can be minimized by selecting a site suitable for a specific proposed activity, rather than by attempting to modify a site to conform to that activity. This kind of planning can be more easily accomplished where there is a general land use plan based on a comprehensive inventory of soils, water, and other related resources.

4.4.2 Determine the Limits of Clearing and Grading. Using a topographic base map that shows existing and finished contours and proposed development and construction, the limits of the areas to be disturbed should be delineated clearly using a heavy line. This line defines the area that should be protected. Regulatory requirements may necessitate identification of high-value waters or other areas warranting special protection, such as steep slopes.

4.4.3 Divide the Site into Natural Drainage Areas. The drainage pattern of a site has two components: overland flow and channel flow. Both are important in erosion control. All existing streams and major drainage ways should be clearly marked. Also, off-site features, such as streams, lakes, wetlands, and other sensitive features that could be impacted and damaged by sediment, should be identified and protected.

4.4.4 Minimize the Area and Duration of Exposure. Another important planning concept is to minimize the extent of the area exposed at one time and the duration of exposure. When land disturbances are required and the natural vegetation is removed, keep the area and the duration of exposure to a minimum. Plan the stages of development so that only the areas that are actively being developed are exposed. All other areas should have a full cover of either temporary or permanent vegetation, or mulch. Grading should be completed as soon as possible. Immediately after grading is completed, a permanent measure, such as vegetative cover, should be established. As cut slopes are made and as fill slopes are brought up to grade, these areas also should be re-vegetated. This is known as staged re-vegetation. Minimizing grading of large or critical areas during the rainy season (time of maximum erosion potential) reduces the risk of erosion.

4.4.5 Apply Perimeter Control Practices. Perimeter control practices should be applied first to protect the disturbed area from off-site runoff and to prevent sedimentation damage to areas below the development site. This concept relates to using practices that effectively isolate the development site from surrounding properties, and especially controls sediment once it is produced, thereby preventing its transport from the site. Diversion, dikes, sediment traps, vegetative filters, and sediment basins are examples of practices to control sediment. Diversion of off-site runoff from the disturbed area should help in reducing the sediment generated and the volume, types, and cost of control practices needed. Vegetative and structural sediment control measures can be classified as either temporary or permanent depending on whether they should remain in use after development is complete. Generally, sediment can be retained by two methods (District of Columbia 2003): (1) filtering runoff as it flows through an area, and (2) impounding the sediment-laden runoff for a period so that the soil particles settle out. The best way to control sediment, however, is to prevent erosion as much as possible.

4.4.6 Apply Erosion Control Practices to Prevent Excessive On-Site Damage. This sixth planning concept relates to using practices that control erosion on a site to prevent excessive sediment from being produced. Soil should be kept covered as much as possible with temporary or permanent vegetation or with various mulch materials. Special grading methods, such as roughening a slope on the contour or tracking with a cleated dozer, may be used. Other practices include diversion structures to direct surface runoff from exposed soil and grade stabilization structures to control erosion. "Gross" erosion in the form of gullies is prevented by these water control devices. Lesser types of erosion, such as sheet and rill erosion, should be prevented when possible. However, often scheduling or the large number of practices required makes this impractical. When erosion is not adequately controlled, sediment control is more difficult and expensive.

4.4.7 Keep Runoff Velocities Low and Retain Runoff On Site. The removal of existing vegetative cover and the increase in impermeable surface area during development can increase both the volume and velocity of runoff. These increases should be taken into account when providing for erosion control. Keeping slope lengths short and gradients low, and preserving natural vegetative cover, can keep stormwater velocities low and limit erosion hazards. Runoff from the development should be conveyed safely to a stable outlet using storm drains, diversions, stable waterways, or similar measures. Consideration should be given to the incorporation of on-site stormwater detention structures to prevent flooding and damage to downstream facilities resulting from increased runoff from the site.

On-site detention may be required by the local jurisdiction. Conveyance systems should be designed to withstand the velocities of projected peak discharges. These facilities should be operational as soon as possible after the start of construction.

4.4.8 Stabilize Disturbed Areas Immediately. Permanent structures, temporary or permanent vegetation, and mulch, or a combination of these measures, should be employed as quickly as possible after the land is disturbed. Temporary vegetation and mulches can be most effective under conditions where it is not practical to establish permanent vegetation. Such temporary measures should be employed immediately after rough grading is completed if a delay is anticipated in obtaining finished grade. The finished slope of a cut or fill should be stable, and ease of maintenance should be considered in the design. Stabilize roadways, parking areas, and paved areas with gravel subbase whenever possible. Many jurisdictions have an upper time limit within which temporary or permanent stabilization should be achieved in areas without active construction.

4.4.9 Implement a Thorough Maintenance and Follow-Up Program. This last concept is vital to the success of the eight other concepts. A site cannot be effectively controlled without thorough, periodic inspection and maintenance of the erosion and sediment control practices. These practices should be maintained just as construction equipment should be maintained and materials checked and inventoried. An example of applying this principle would be to start a routine "end-of-day check" to make sure that all control practices are working properly.

4.4.10 Summary of Concepts. Usually, these nine concepts are integrated into a system of vegetative measures and structural measures along with management techniques to develop a plan to prevent erosion and control sediment. In most cases, a combination of limited grading, limited time of exposure, and a judicious selection of erosion control practices and sediment trapping facilities should prove to be the most practical method of controlling erosion and the associated production and transport of sediment.

CHAPTER 5
CONTROL CONCEPTS, PRACTICES, AND STANDARDS

This chapter provides a brief overview of the concepts used in planning, implementing, and maintaining an erosion and sediment control system, including identification of the most commonly used practices.

5.1 EROSION AND SEDIMENT CONTROL CONCEPTS

In planning, implementing, and maintaining an erosion and sediment control system, it is important to understand the difference between erosion control and sediment control. An effective erosion and sediment control plan should consist of a combination of these important concepts.

5.1.1 Erosion Control Practices. Erosion control practices (source protection) are a group of activities that can be used to protect the soil surface and prevent soil particles from being detached by rainfall or wind. Erosion control can be viewed as a source control that treats the soil as a resource that has value and should be kept in place. Erosion prevention practices should be grouped into practices to control graded areas, slopes, and drainage ways (Krause et al. 2008).

5.1.2 Sediment Control Practices. Sediment control practices are a group of techniques that can be used to trap the soil particles after they have been detached and moved by wind or water. Sediment control measures are usually passive systems that rely on filtering or settling the particles out of the water or wind that is transporting them. Sediment control treats the soil as a waste product that should be removed from where it has been transported and accumulated and disposed of at another location. Sediment control measures are often grouped into the categories as provided in Section 5.3.

5.1.3 Operation and Maintenance. Operation and maintenance includes periodic inspections, performed after storms and after repairs. Also, as controls are no longer needed, these controls should be removed and areas should be revegetated as appropriate. Sediment removed from erosion controls as part of maintenance procedures should be stored in a suitable area and in a manner that will not erode; eventually this can be incorporated into the landscape and seeded, assuming the sediments are not considered to be contaminated or hazardous. Controls where sediment removal was conducted should be restored to original dimensions. In general, projects should keep the point of discharge free of erosion.

As there is significant disturbance of soil on a construction and development project, managing fill material is an essential component of erosion and sediment control in construction practices and in construction and maintenance of sediment control measures. Compaction of fills should be considered, as appropriate, to reduce erosion, slippage, settlement, subsidence, or other related problems; generally, fill should be compacted with on-site vehicles. Compaction of heavy clay soils should be avoided (USEPA 2006). Note that fill intended to support buildings, structures, conduits, and so on should be compacted in accordance with all geotechnical requirements, local requirements, and codes similar to those in Appendix A. Fill should be placed and compacted in layers not to exceed 200 mm (8 in.) in thickness. Fill should be free of brush, rubbish, hazardous material, rocks, logs, stumps, building debris, and other objectionable materials that would interfere with or prevent construction of satisfactory fills. Frozen material or soft, mucky, or highly compressible materials should not be incorporated into fill slopes or structural fills, and fills cannot be placed on a frozen foundation (MDE 1994a).

5.2 EROSION CONTROL PRACTICES

Erosion control practices can be structured into four major groupings that include (1) site preparation practices and stabilization practices to control graded areas, (2) slopes, (3) roadways, and (4) drainage ways to minimize the generation of sediment from these sensitive areas.

5.2.1 Site Preparation. Site preparation practices that can be used to reduce and prevent excessive generation of sediment from a construction site include (1) land grading and roughening, and (2) topsoiling (District of Columbia 2003).

5.2.1.1 Land Grading/Roughening. Land grading and roughening consists of reshaping of the existing land surface in accordance with a plan as determined by engineering survey and layout. The purpose of a land grading practice is to provide for erosion control and vegetative establishment on those areas where the existing land surface is to be reshaped by grading according to plan.

The grading plan should be based on the incorporation of building designs and street layouts that fit and utilize existing topography and desirable natural surroundings to avoid extreme grade modifications. Most grading plans include information that provides sufficient topographic surveys and soil investigations to determine limitations that should be imposed on the grading operation related to slope stability, effect on adjacent properties and drainage patterns, measures for drainage, water removal and vegetative treatment, and others.

The grading plan should show existing and proposed contours of the areas to be graded. The plan also should include practices for erosion control, slope stabilization, safe disposal of runoff water and drainage, such as waterways, lined ditches, reverse slope benches (including grade and cross section), grade stabilization structures, retaining walls, and surface and subsurface drains. The plan should include phasing of these practices as well. Fig. 5-1 shows typical details in design of a reverse slope bench.

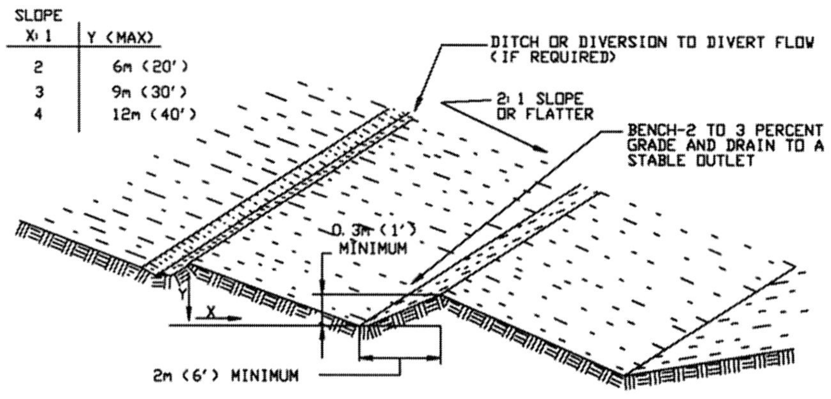

FIGURE 5-1. Reverse Slope Bench
***Source*: Fig. F193, Detail 28, benched slopes (MDE 1994b), with modifications by MPCES Committee, reproduced with permission from the Maryland Department of the Environment.**

The following are additional construction specifications for Fig. 5-1:

- All benches should be kept free of sediment during all phases of development. All graded areas should be permanently stabilized immediately following finished grading.
- Seeps or springs encountered during construction should be handled in accordance with the National Engineering Handbook (USDA/NRCS 2001a) or other approved methods.

5.2.1.2 Topsoiling. Topsoiling consists of placement of topsoil over prepared subsoil prior to establishment of permanent vegetation to provide a suitable soil medium for vegetative growth. Soils benefiting from topsoiling have low moisture content, low nutrient levels, non-optimum pH, materials toxic to plants, and/or unacceptable soil gradation. The topsoiling practice typically is limited to areas having 2:1 or flatter slopes (District of Columbia 2003).

Other considerations are that the texture of the exposed subsoil/parent material is not adequate to produce vegetative growth, soil material is so shallow that the rooting zone is not deep enough to support plants or furnish continuing supplies of moisture and plant nutrients, and/or the original soil to be vegetated contains material toxic to plant growth. In addition, the soil may be so acidic that treatment with limestone is not feasible.

5.2.2 Vegetative Stabilization. Vegetative stabilization consists of using vegetation as cover for barren soil to protect it from forces that cause erosion. Vegetative stabilization practices are used to promote the establishment of vegetation on exposed soil. When soil is stabilized with vegetation, the soil is less likely to erode and more likely to allow infiltration of rainfall, thereby reducing sediment loads and runoff to downstream areas, and improving wildlife habitat and visual resources.

This practice is used on denuded areas as specified on the plans and may be used on highly erodible or critically eroding areas. This practice is divided into (1) temporary seeding to quickly establish vegetative cover for a short duration (up to 1 year), and (2) permanent seeding for long-term vegetative cover. Examples of applicable areas for temporary seeding are temporary soil stockpiles, cleared areas being left idle between construction phases, earth dikes, and the like; and for permanent seeding, lawns, dams, cut and fill slopes, and other areas at final grade, former stockpile, and staging areas, among others (District of Columbia 2003).

Planting vegetation in disturbed areas will have an effect on the water budget, especially on volumes and rates of runoff, infiltration, evaporation, transpiration, percolation, and groundwater recharge. Vegetation, over time, will increase organic matter content and improve the water-holding capacity of the soil and subsequent plant growth.

Vegetation will help reduce the movement of sediment, nutrients, and other chemicals carried by runoff to receiving waters. Plants also help protect groundwater supplies by assimilating those substances present within the root zone.

Sediment control devices should remain in place during grading, seedbed preparation, seeding, mulching, and vegetative establishment to prevent large quantities of sediment and associated chemicals and nutrients from washing into surface waters.

5.2.3 Blankets and Matting. Blankets and matting consists of the installation of a protective covering (blanket) or a soil stabilization mat on a prepared planting area of a steep slope, channel or shoreline. The purpose of this practice is to aid in controlling erosion on critical areas by providing a microclimate that protects young vegetation and promotes its establishment. Note that care should be exercised in blanket/matting placement to prevent rapid erosion of soil under undeveloped vegetation, particularly at the mat-soil interface (Pitt et al. 2007). In addition, some types of soil stabilization mats are used to raise the maximum permissible velocity of turf grass stands in channelized areas by "reinforcing the turf" to resist the forces of erosion during storm events.

This practice typically is used on short, steep slopes where erosion hazard is high and planting is likely to be too slow in providing adequate protective cover, in vegetated channels where the velocity of design flow exceeds "allowable" velocity, on streambanks or tidal shorelines where moving water is likely to wash out new plantings, or in areas where the forces of wind prevent standard mulching practices from remaining in place until vegetation becomes established.

Soil stabilization blankets and mats can be applied to problem areas to supplement nature's erosion control system (vegetation) in its initial establishment and in providing a safe and "natural" conveyance for high-velocity stormwater runoff. They are being used today in many applications where previously a structural lining would have been required. Care should be taken to choose the type of blanket or matting that promotes the establishment of young vegetation and protects it. A wide range of soil stabilization products are available today, and new products are being introduced. Rafter (2004) indicated how "contractors find new erosion control applications and that the materials are not new. Engineers, developers, and landscapers have been using them for years to help eliminate erosion, stabilize the earth surrounding construction projects, and catch large chunks of sediment that might otherwise escape into streams, rivers, and other bodies of water."

Fig. 5-2 shows blanket applications for shallow slopes, steep slopes, and ditches. On shallow slopes, strips of netting coverings can be applied across the slope. Where there is a berm at the top of the slope, bring the blanket over the berm and anchor behind the berm. On steep slopes, apply the blanket parallel to the direction of flow and anchor securely. The blanket should be brought down to a level area before terminating the installation. The ends of the blanket should be turned under a minimum of 100 mm (4 in.) and stapled at 300 mm (12 in.) intervals. In ditches, the blanket should be laid parallel to the direction of flow and check slots used as required; overlapping joints should be placed parallel in the direction of flow. Blankets should not be joined in the center of the ditch. Fig. 5-3 provides typical details for generic mats used for slope stabilization.

5.2.4 Road Stabilization. Areas that are graded for construction vehicle transport and parking purposes are especially susceptible to erosion. The exposed soil surface is continually disturbed, leaving no opportunity for vegetative stabilization. Such areas also tend to collect and transport runoff water along their surfaces. During wet weather, they often become muddy quagmires, which generate significant quantities of sediment that may pollute nearby streams or be transported off site on the wheels of construction vehicles. Dirt roads can become so unstable during wet weather that they are virtually unusable.

Permanent roads and parking areas should be paved as soon as possible after grading. However, it is understandable that weather conditions or the potential for damage may not make paving feasible in the early phases of the development project. As an alternative, the early application of stone may solve potential erosion and stability problems and eliminate later regrading costs. Some of the stone should also remain in place for use as part of the final base course in the construction of the road.

Three individual prevention practices are available to address road construction issues. They include construction road stabilization, stabilized construction entrances, and stabilized construction entrances with wash racks (District of Columbia 2003).

5.2.4.1 Construction Road Stabilization. Construction road stabilization consists of the temporary stabilization of access

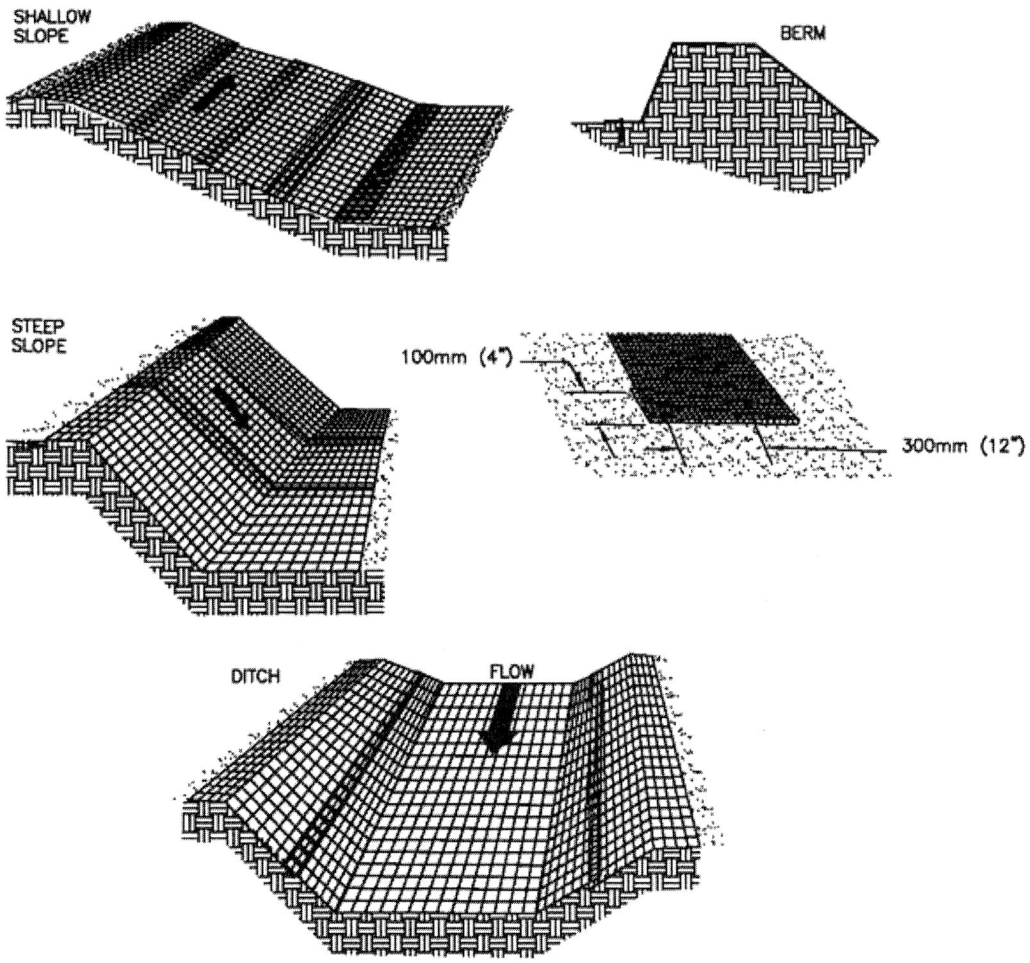

FIGURE 5-2. Soil Stabilization Blanket

Source: Plate 3.36-1, Chapter 3, typical orientation of treatment–1, soil stabilization blanket (VDEQ 1992b), with modifications by MPCES Committee, reproduced with permission from the Virginia Department of Environmental Quality.

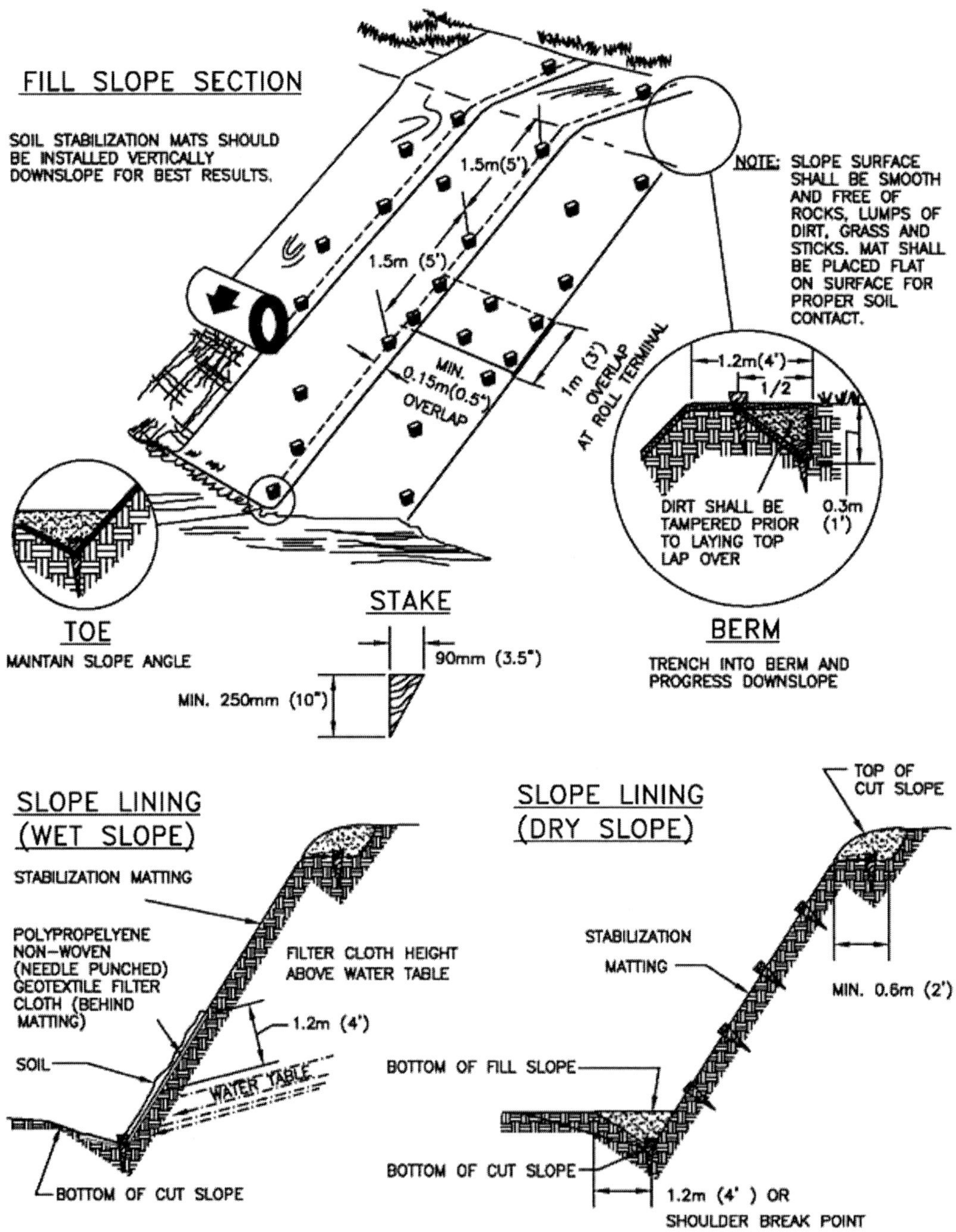

FIGURE 5-3. Slope Stabilization with Mat

***Source*: Plate 3.36-5, Chapter 3, typical treatment–2, soil stabilization matting slope installation (VDEQ 1992c), with modifications by MPCES Committee, reproduced with permission from the Virginia Department of Environmental Quality.**

roads, haul roads, temporary construction parking areas and other on-site vehicle transportation routes with stone immediately after grading.

5.2.4.2 Stabilized Construction Entrances. A stabilized construction entrance consists of a stabilized layer of aggregate that is underlain with geotextile fabric (District of Columbia 2003). Stabilized entrances are located at any point where traffic enters or leaves a construction site. Vehicles leaving the site should travel over the entire length of the stabilized construction entrance. Fig. 5-4 provides typical details of a stabilized construction entrance.

The stabilized construction entrance should be flared at the existing road to provide a turning radius. Nonwoven geotextile fabric should be placed over the existing ground prior to placing stone to prevent subsidence of the gravel; recycled concrete equivalent (free of rebar) can often be used instead of crushed stone (MDE 2011). All surface flowing water to or diverted toward construction entrances should be piped through the entrance, maintaining positive drainage, and the

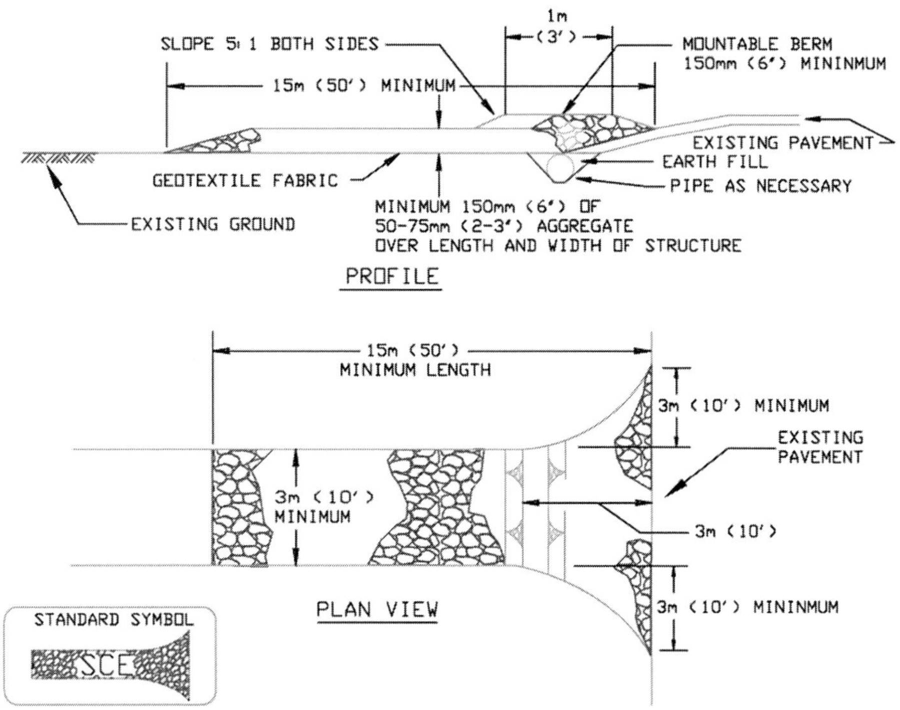

FIGURE 5-4. Stabilized Construction Entrance (SCE)
***Source*: Fig. F173, Detail 24, stabilized construction entrance (MDE 1994c), with modifications by MPCES Committee, reproduced with permission from the Maryland Department of the Environment.**

pipe installed through the stabilized construction entrance should be protected with a mountable berm with 5:1 slopes and a minimum of 150 mm (6 in.) of stone over the pipe. Pipe should be sized according to the amount of runoff to be conveyed, though 150 mm (6 in.) minimum is recommended (MDE 1994c). When possible the stabilized construction entrance should be located at a high spot, which may not require a pipe.

5.2.4.3 Stabilized Construction Entrances with Wash Racks. A stabilized construction entrance with wash rack consists of a stabilized layer of aggregate, which is underlain with geotextile fabric and enhanced by the use of a wash rack (District of Columbia 2003). Construction vehicles exiting the site should first stop with first the front wheels and then the rear wheels over the wash rack so that both sets of wheels can be hosed off before driving off to the existing roadway. Fig. 5-5 provides typical details of a stabilized construction entrance with a wash rack.

The stabilized construction entrance thickness requires constant maintenance to retain the specified dimensions, so a stockpile of rock material is required for adding rock to maintain effectiveness (PADEP 2012). In addition, the drain space under wash rack should be kept open at all times.

5.3 SEDIMENT CONTROL MEASURES

Sediment control measures are employed to intercept, redirect, or amass mobilized sediment. They include sediment barriers, traps, and basins.

5.3.1 Sediment Barriers. Pollution by sediment is a primary carrier of sorbed chemicals, especially phosphorus, chlorinated pesticides, and metals, which are transported by sediment into aquatic systems (FAO 1996). Woelkers and Theisen (2003) have indicated, "Appropriate places to use sediment control barriers include along sections of a site perimeter, below disturbed areas subject to sheet and rill erosion, etc." This reference also provides useful tables about many different BMPs grouped by *function* rather than by construction phases of projects.

Sediment barriers include a series of temporary barriers used to intercept and direct sediment-laden runoff. The measures include silt fences, super silt fences, storm drain inlet protection, inlet filter bags, and culvert inlet protection (District of Columbia 2003).

Beecham (2010) suggests using wattles for sediment control. Another author, Lanka Santha, P.E., suggested the use of "logs and wattles," as included within the article by Northcutt (2008), who referenced many professionals, stating, "Properly selected and installed products can help reduce the cost of trapping sediment during and after solid-disturbing activities."

Silt fences are a temporary barrier of geotextile fabric used to intercept and reduce velocity in disturbed areas. Silt fences intercept sediment-laden runoff so that deposition of the transported sediment can occur. Use of a silt fence should be limited to intercepting sheet flow only from short-sloped flow paths. Silt fences generally should be placed parallel to the slope and perpendicular to flow. Silt fences should not be used as velocity checks in ditches or swales, or placed where they will intercept concentrated flow. Fig. 5-6 provides typical specifications for a silt fence installation.

Additional construction specifications for Fig. 5-6 include using quality hardwood for wooden posts; square posts should be minimal dimensional 2 × 2 lumber while round posts should be minimum 44 mm (1.75 in.) diameter (MDE 1994d). Steel posts can be substituted but should be standard T or U section weighing not less than 1.5 kg/m (1 lb/ft). Geotextile should be fastened securely to each fence post with wire ties or staples at

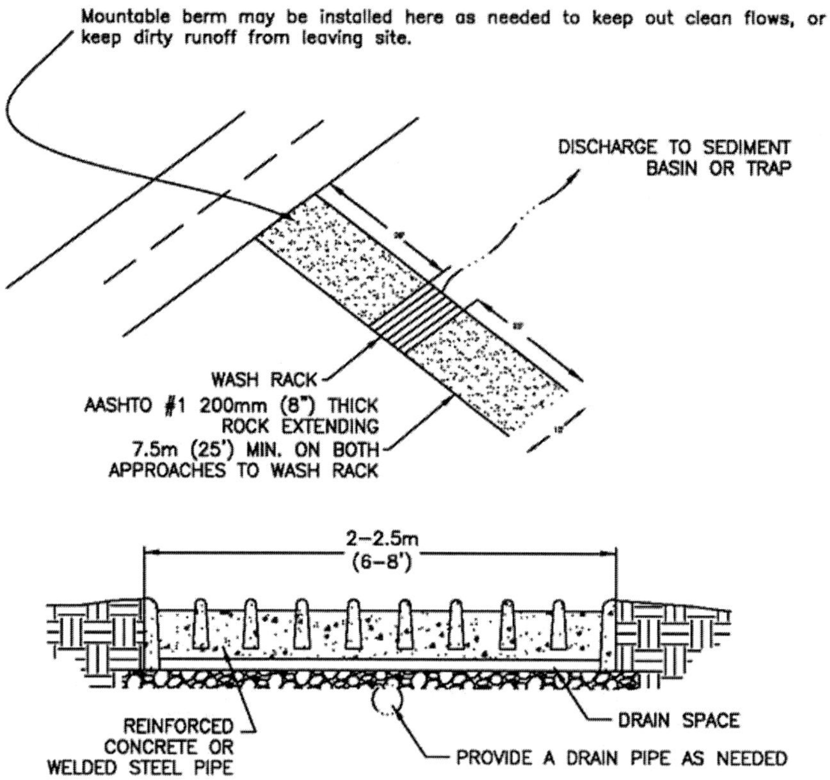

FIGURE 5-5. Stabilized Construction Entrance with Wash Rack

***Source*: Standard construction, Detail 3-2, rock construction entrance with wash rack (PDEP 2012), with modifications by MPCES Committee, reproduced with permission from the Pennsylvania Department of Environmental Protection.**

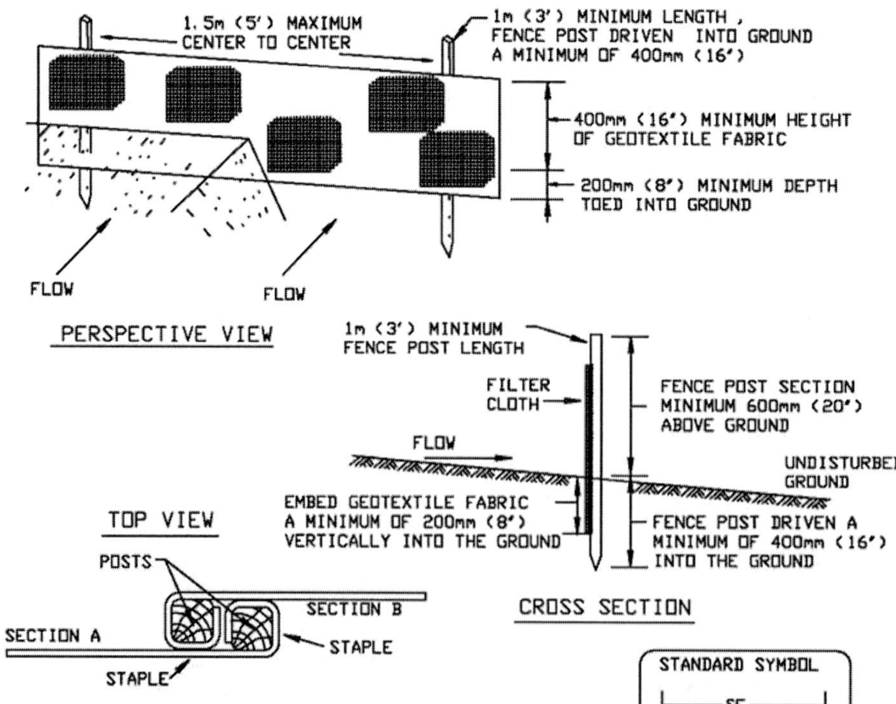

FIGURE 5-6. Silt Fence

***Source*: Fig. E153, Detail 22, silt fence (MDE 1994d), with modifications by MPCES Committee, reproduced with permission from the Maryland Department of the Environment.**

top and midsection. Where ends of geotextile fabric come together, they should be overlapped, folded, and stapled to prevent sediment bypass as shown in detail. Remove sediment when bulges occur or when accumulation reaches 30% of the fabric height.

Geotextile fabrics, woven or nonwoven (spun), used in the construction of silt fences and other exposed settings (above subgrade) should resist deterioration from solar radiation (ultraviolet) and be sturdy enough for a minimum period of 12 months of expected usable construction life at a temperature range of −18 to 49°C (0 to 120°F) (MDE 1994d). Barrett et al. (1998) noted that of several geotextiles tested, the geotextile fabric with an equivalent opening size (EOS) of 150 μm (0.000039 in.) had the longest detention time, which improved sedimentation behind the silt fence; however, local regulations may require greater openings, which may be specified by sieve number. Other common specification criteria include tensile strength and elongation (ASTM 2008); weathering and ultraviolet stability (ASTM 2007) may be specified based on percent retention of tensile strength. In general, woven geotextiles should be used perpendicular to flow (e.g., silt fences and straw bale dikes), whereas nonwoven, needle-punched fabric, typically every 150 to 300 mm [6 to 12 in.] are primarily used as geotextile barriers to separation between stone layers and earth. If used for filtering purposes, replace the geotextile fabric when it becomes clogged.

Super silt fences are temporary barriers of geotextile fabric over chain-link fence used to intercept sediment-laden sheet flow from drainage areas (District of Columbia 2003). The super silt fence reduces the runoff velocity and allows the deposition of transported sediment to occur. Super silt fences should not be used as velocity checks in ditches or swales or placed where they will intercept concentrated flow. A super silt fence may provide a barrier to prevent sediment and debris from entering critical areas, such as streets and streams. Fig. 5-7 provides typical details for a super silt fence installation.

Securely fasten geotextile filter cloth to the chain-link fence with ties spaced every 0.6 m (2 ft) at the top and midsection while the lower tension wire, brace and truss rods, drive anchors, and post caps are typically only required on the ends of the fence (MDE 1994e). Overlap two sections of filter cloth as demonstrated in Fig. 5-7. Perform maintenance when "bulges" develop in the silt fence or when silt reaches 30% of fence height (MDE 1994e).

Storm drain inlet protection is a temporary barrier constructed around a storm drain inlet. The storm drain inlet protection filters and traps the sediment before it can enter the storm drain system. The storm drain inlet protection is a secondary sediment control device and should not be used in place of a sediment trapping device. Fig. 5-8 provides typical details for storm drain inlet protection.

Additional construction specifications for Fig. 5-8 include driving 2 × 4 posts (construction grade lumber) 0.3 m (1 ft) into the ground at each corner of the inlet, with nail strips between the posts (MDE 1994f). Assemble the top portion of the 2 × 4 frame using the overlap joint shown in Fig. 5-8. The frame (weir) should be a minimum of 0.15 m (0.5 in.) below adjacent roadways where flooding and safety issues may arise. Stretch a wire mesh tightly around the frame and fasten securely, overlapping the ends at a post. The suggested maximum wire mesh aperture is 13 × 13 mm (0.5 × 0.5 in.). Stretch the geotextile fabric tightly over the wire mesh with the geotextile extending from the top of the frame to 0.45 m (1.5 ft) below the inlet notch elevation. The ends of the geotextile should meet at a post, be overlapped and folded, then fastened firmly to the frame. Backfill around the inlet in compacted 0.15 m (0.5 ft) layers until the layer of earth is level with the notch elevation on the ends and top elevation on the sides. Construct a compacted earth dike across the ditch line directly below it only if the inlet does not contain a sump. The top of this earth dike should be at least 0.15 m (0.5 in.) higher than the top of the frame. Replace the geotextile fabric when it becomes clogged.

Inlet filter bags are temporary barriers placed inside storm drain inlets to filter and trap sediment and, thus, keep it out of the storm drain system. This measure is not recommended on finished roadways where ponding can cause traffic hazards, and a maximum drainage area of 0.2 ha (0.5 ac) is recommended. Fig. 5-9 provides a typical detail for installing an inlet filter bag.

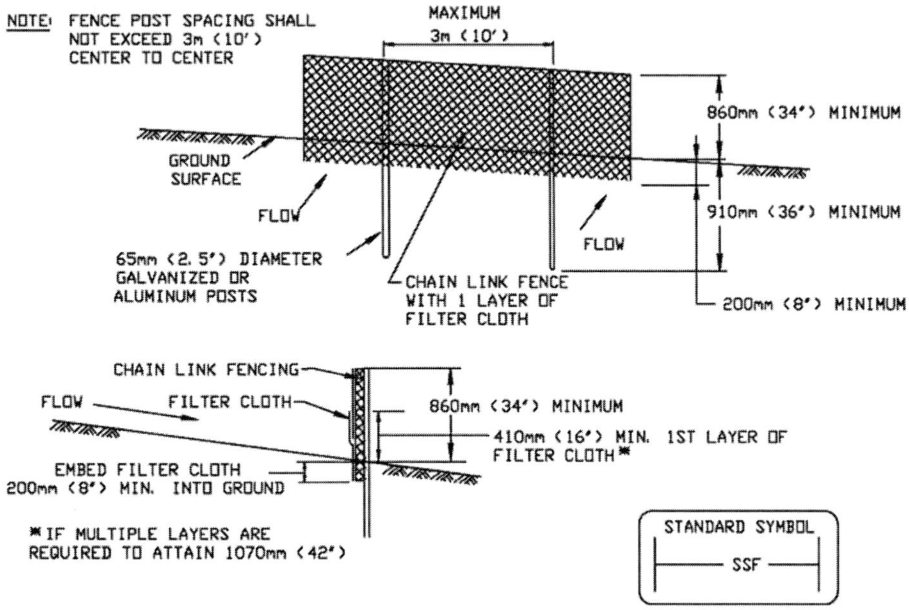

FIGURE 5-7. Super Silt Fence

***Source*: Fig. H263, Detail 33, super silt fence (MDE 1994e), with modifications by MPCES Committee, reproduced with permission from the Maryland Department of the Environment.**

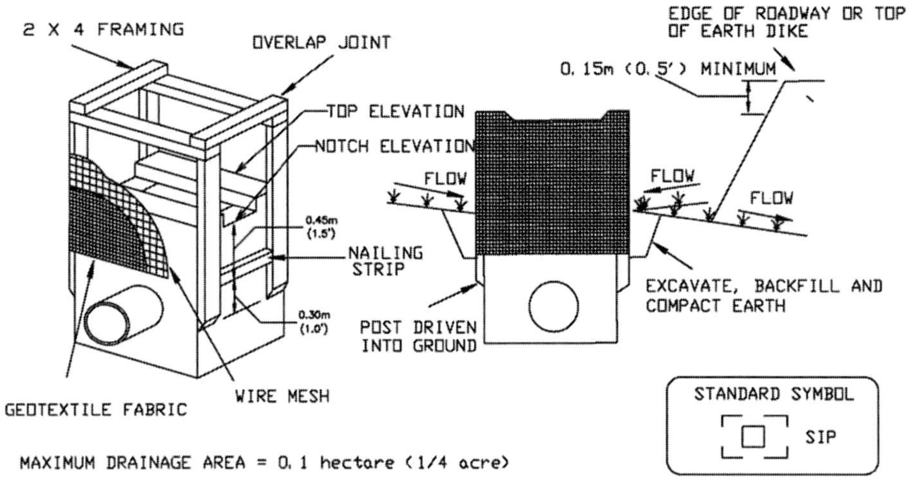

FIGURE 5-8. Inlet Protection
***Source*: Fig. E165, Detail 23A, standard inlet protection (MDE 1994f), with modifications by MPCES Committee, reproduced with permission from the Maryland Department of the Environment.**

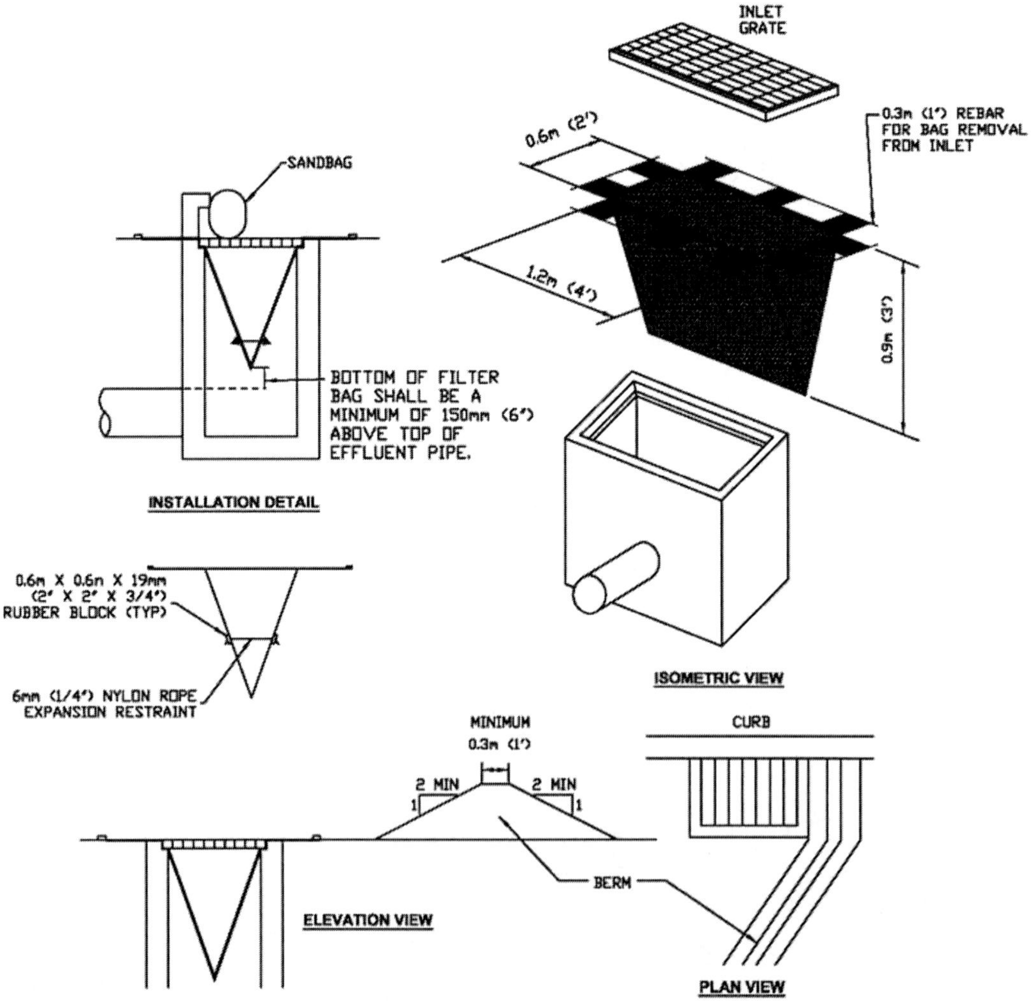

FIGURE 5-9. Inlet Filter Bag
***Source*: Standard construction, Detail 28, filter bag inlet protection curbed roadway (PDEP 2000), with modifications by MPCES Committee, reproduced with permission from the Pennsylvania Department of Environmental Protection.**

Typically, berms are required for all installations, starting with an earthen berm until a roadway is stoned, progressing to a road subbase berm until a roadway is paved, and finally an asphalt berm until a roadway surface receives a final coat.

Culvert inlet protection consists of a temporary sediment filter located at the inlet to a culvert. This barrier is used to prevent sediment from entering, accumulating in, and being transferred by a culvert and associated drainage system prior to permanent stabilization of a disturbed area. The silt fence is placed a minimum of 2 m (6 ft) from the culvert, encircling the inlet. Fig. 5-10 provides a typical detail for a culvert inlet protection.

5.3.2 Sediment Traps. Sediment traps are temporary sediment control devices formed by excavation and/or an embankment with an approved outlet. The purpose of the sediment trap is to intercept sediment-laden runoff and trap the sediment in order to protect drainage ways, properties, and rights-of-way downstream of the sediment trap from sedimentation. One example is Fig. 5-11.

The coarse aggregate and stone riprap are laid out in a horseshoe shape around the inlet. The maximum drainage area for this measure is 1.2 ha (3 ac) with a sediment trap volume to drainage area of 125 m^3/ha (1800 ft^3/ac) for both the wet (below base of the stone) and dry (from base of the stone) storage. Maintenance by clean-out of the sediment trap is to occur when half of the wet stage depth is filled in.

There are four types of sediment traps, which are differentiated by the type of outlet structure. They include

1. Pipe outlet sediment trap
2. Stone outlet sediment trap
3. Riprap outlet sediment trap
4. Stone/riprap outlet sediment trap.

To use any of these measures, clear, grub, and strip away any vegetation and root mat from the area under the embankment and size pool or wet storage area according to drainage area specifications. Fill material for the embankment should be free of roots or other woody vegetation, as well as oversized stones, rocks, organic material, or other objectionable material. Compact the embankment by traversing with construction equipment during construction.

A pipe outlet sediment trap is a sediment trap that uses a pipe outlet device. The outlet is through a perforated riser and a barrel pipe through the embankment. The barrel pipe and riser can be made of corrugated metal or PVC. The pipe outlet sediment trap can be used to control a drainage area up to 2 ha (5 ac) with a combined wet and dry storage to drainage area defined as 250 m^3/ha (3600 ft^3/ac). The top of embankment should be a minimum of 0.3 m (1 in.) above the riser crest elevation. Fig. 5-12 provides typical details for a pipe outlet sediment trap.

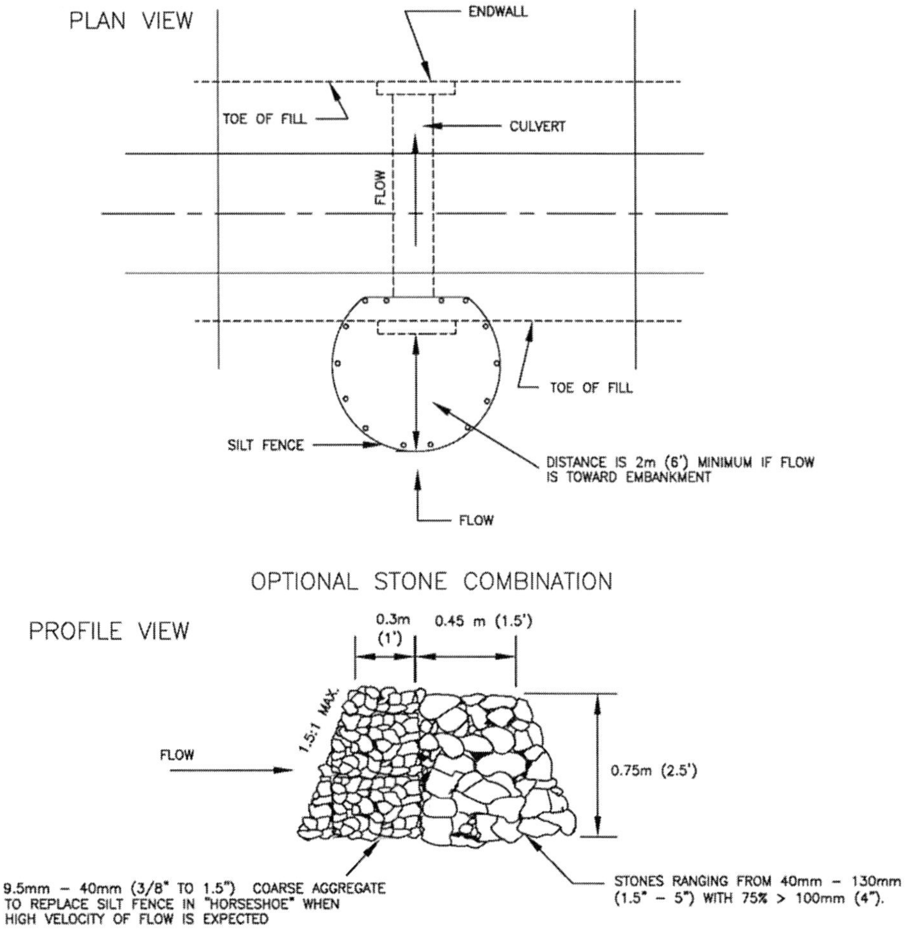

FIGURE 5-10. Silt Fence Culvert Protection

***Source*: Plate 3.08-1, Silt fence culvert inlet protection (VDEQ 1992d), with modifications by MPCES Committee, reproduced with permission from the Virginia Department of Environmental Quality.**

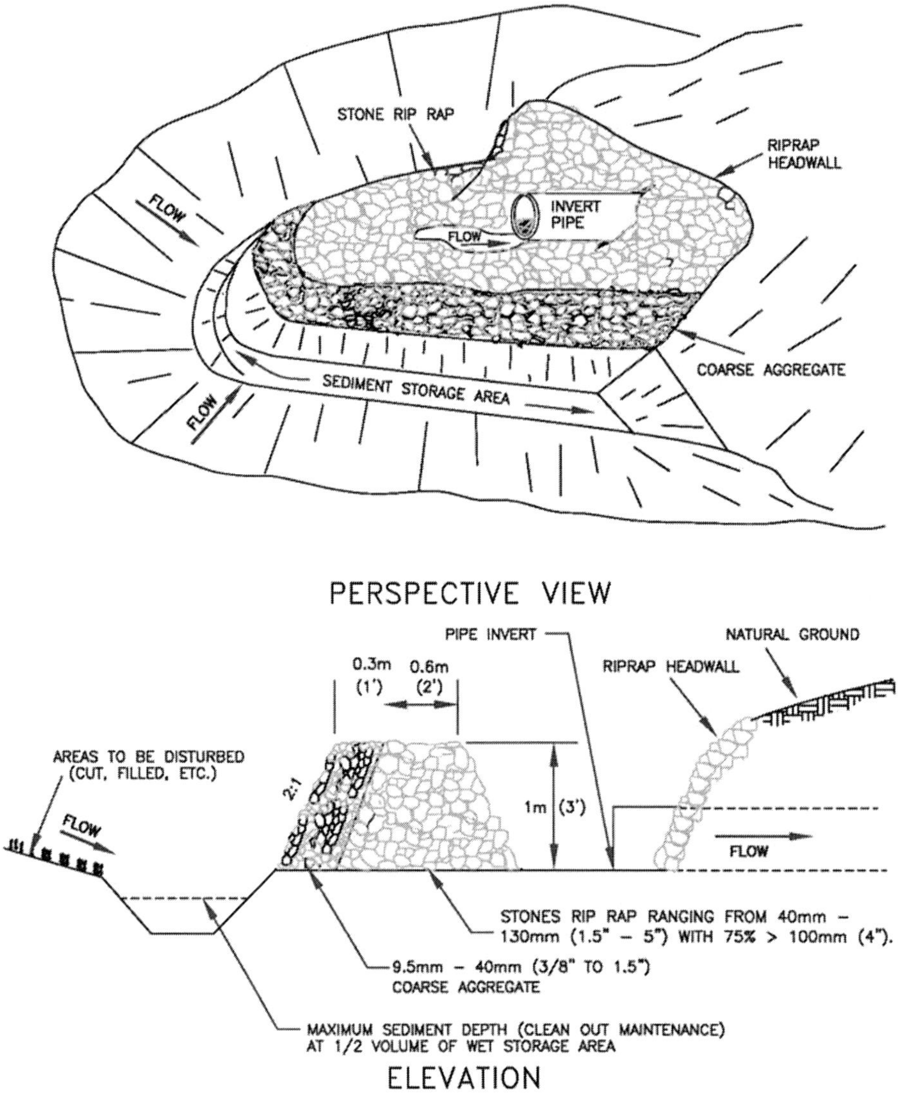

FIGURE 5-11. Stone Culvert Inlet Protection with Sediment Trap
***Source*: Plate 3.08-2, Culvert inlet sediment trap (VDEQ 1992e), with modifications by MPCES Committee, reproduced with permission from the Virginia Department of Environmental Quality.**

In addition, construction specifications for Fig. 5.12 are to "perform maintenance for sediment removal when the sediment has accumulated to one half of the wet storage depth of the trap; this is 62.5 m^3/ha (900 ft^3/ac) or 0.3 m (1 ft) if built to maximum fill height of 1.2 m (4 ft)" (MDE 1994g).

A stone outlet sediment trap is a trap that uses a stone outlet device. The outlet for this trap is over a stone section placed on level ground. The minimum weir length should be 3 m/ha (4 ft/ac) of the drainage area. The stone outlet sediment trap can be used to control a drainage area of up to 2 ha (5 ac) in size. Fig. 5-13 provides a typical detail for a stone outlet sediment trap.

An additional construction specification for Fig. 5-13 is that geotextile fabric may be substituted for the stone facing by placing it on the upstream face of the stone outlet (MDE 1994h).

A riprap outlet sediment trap is a trap that uses a riprap outlet device. The outlet for this trap is through a partially excavated channel lined with riprap or recycled concrete equivalent. The outlet channel should discharge onto a stabilized area or to a stable watercourse. The riprap outlet sediment trap can be used to control a drainage area up to 4 ha (10 ac) in size. Fig. 5-14 provides a typical detail for a riprap outlet sediment trap.

A stone/riprap outlet sediment trap is a trap that uses a combination of a stone and riprap outlet device. The outlet for this trap is through a partially excavated channel lined with riprap and containing a stone or recycled concrete equivalent outlet structure. The stone/riprap outlet sediment trap can be used to control a drainage area up to 4 ha (10 ac) in size. The outlet should discharge onto a stabilized area or to a stable watercourse with a minimum 1.5 m (5 ft) apron for up to 2 ha (5 ac) drainage area and 3 m (10 ft) for 2 to 4 ha (5 to 10 ac). Fig. 5-15 provides a typical detail for a stone/riprap outlet sediment trap.

The riprap outlet should be constructed to full course thickness in one operation, avoiding displacement of underlying materials. Place stone for riprap outlet in a manner to ensure homogeneity, with the smaller stones and spalls filling the voids between the larger stones, and prevent damage to the stone filter blanket or geotextile (hand place to the extent necessary) (MDE 2011).

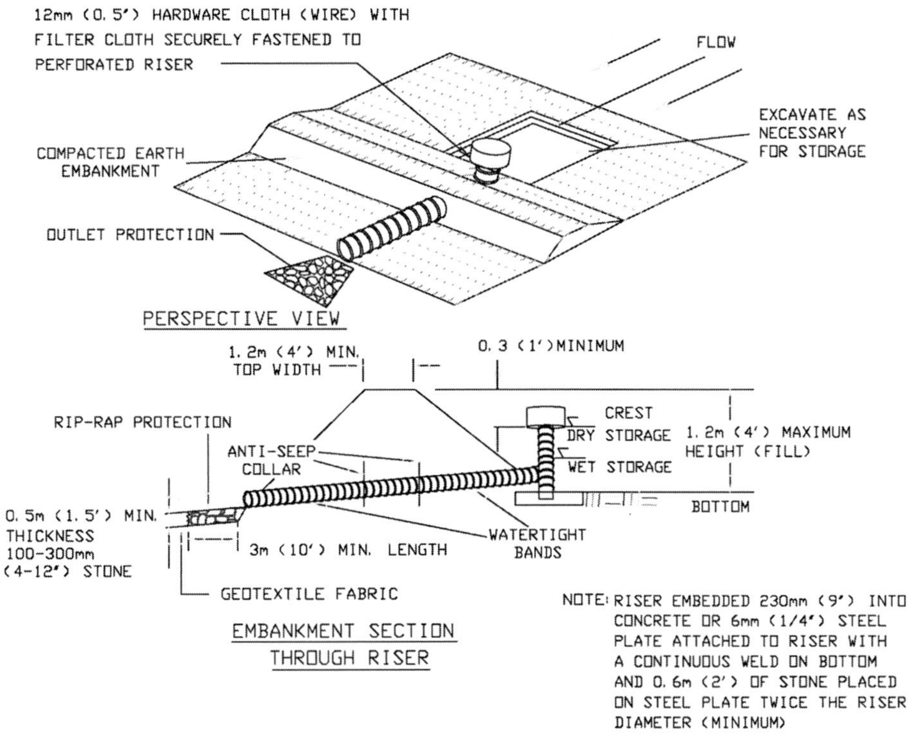

FIGURE 5-12. Pipe Outlet Sediment Trap

***Source*: Fig. C97, Detail 8, Pipe outlet sediment trap-ST I (MDE 1994g), with modifications by MPCES Committee, reproduced with permission from the Maryland Department of the Environment.**

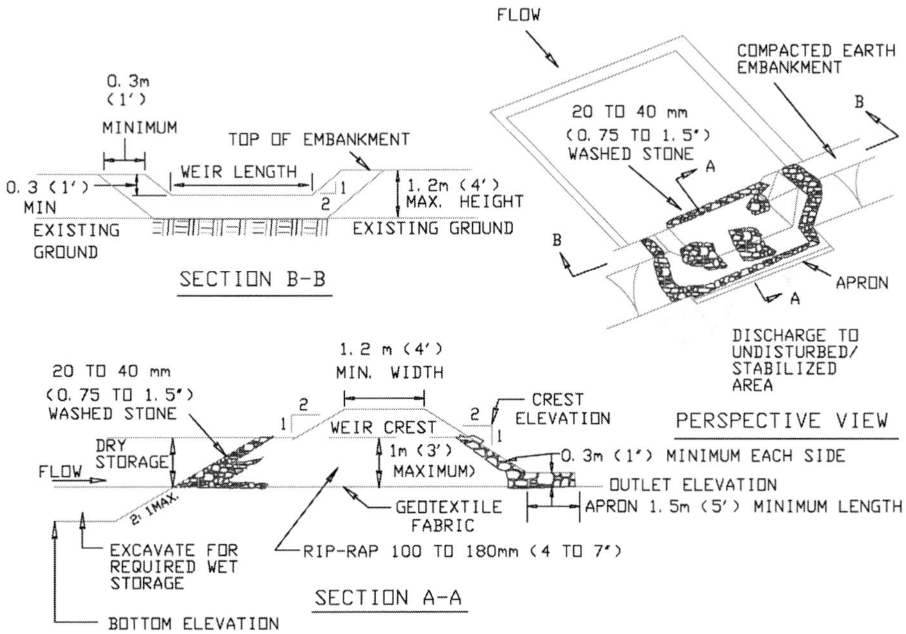

FIGURE 5-13. Stone Outlet Sediment Trap

***Source*: Fig. C910, Detail 9, Stone outlet sediment trap (ST-2) (MDE 1994h), with modifications by MPCES Committee, reproduced with permission from the Maryland Department of the Environment.**

5.3.3 Sediment Basins. A sediment basin is a temporary barrier or dam constructed across a drainage way (as opposed to a waterway) or at other suitable locations to intercept sediment-laden runoff. The design of a sediment basin may incorporate a combination of a sized barrier/dam with excavation to achieve the required storage. Sediment basins protect downstream properties and drainage ways by trapping sediment and controlling the release of stormwater runoff.

A sediment basin should be required to control runoff and sediment from large areas, typically greater than 4 ha (10 ac),

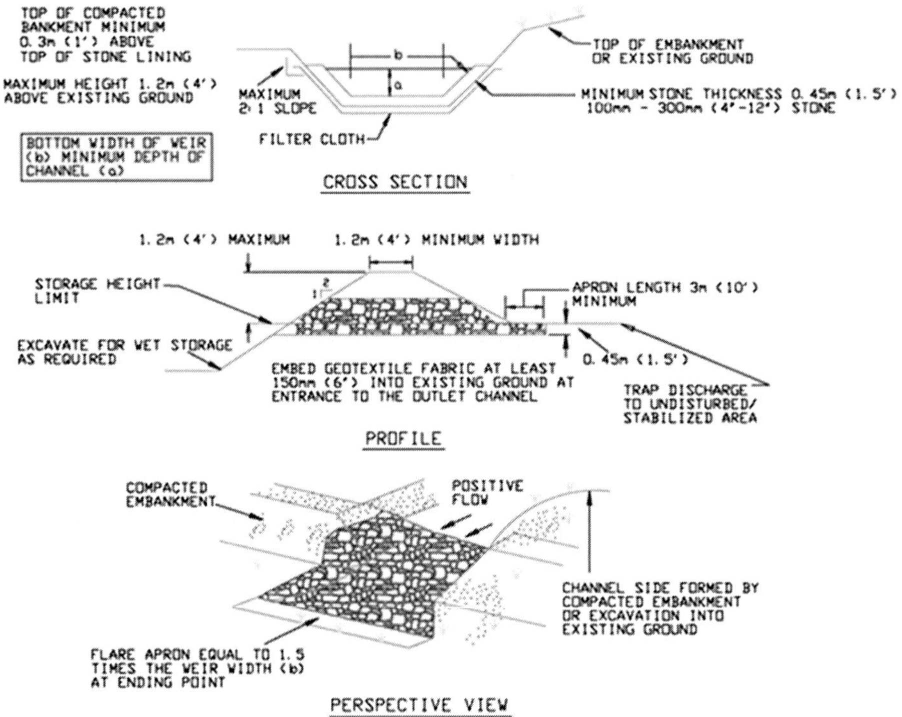

FIGURE 5-14. Riprap Outlet Sediment Trap

***Source*: Fig. C913, Detail 10, Riprap outlet sediment trap (ST-III) (MDE 1994i), with modifications by MPCES Committee, reproduced with permission from the Maryland Department of the Environment.**

where sediment traps are not appropriate. Stormwater management ponds may be used as sediment basins provided that they meet the requirements of the sediment basin and that the construction sequence addresses converting the sediment basin to the permanent stormwater management pond. Fig. 5-16 provides typical details for a sediment basin.

Representative dimensions beyond recommended minimums have not been provided in Fig. 5-16, as there is a significant amount of engineering design considerations and site-specific details required to design a sediment basin. Refer to local regulations and design guides concerning basin design. Maximum drainage area to a stormwater basin is typically 40 ha (100 ac).

5.4 WATER CONVEYANCE MEASURES

A number of control measures can be used to intercept and safely convey concentrated stormwater flows on construction sites without causing accelerated erosion and sediment generation. These measures can be grouped as (1) dikes and diversions, (2) downdrains and flumes, and (3) grass or lined waterways (District of Columbia 2003).

5.4.1 Dikes and Diversions. Dikes and diversions is a group of measures that include earth dikes, temporary swales, dike/swales, and temporary storm drain diversions (District of Columbia 2003).

Earth dikes are temporary berms or ridges of soil, compacted, stabilized, and located in such a manner as to direct water to an approved location. The purpose of the earth dike is to direct runoff to a sediment trapping device, which reduces the potential for erosion and sedimentation. Earth dikes also can be used for diverting clean, off-site runoff water away from disturbed areas. Temporary earth dikes require an uninterrupted positive grade, minimum slope of 0.5%, to an outlet (MDE 1994k).

Earth dikes often are constructed across disturbed areas and around construction sites, such as parking lots and subdivisions. Excavate or shape the dike to line, grade, and cross section as required. The dike should be free of bank projections or other irregularities that would impede normal flow. The dikes should remain in place until the disturbed areas are permanently stabilized. Fig. 5-17 provides a typical detail for an earth dike. Select size of earth dike (letter) based on channel stabilization criteria (number) in Table 5-1; slope and drainage are in Table 5-2 (letter-number).

Temporary swales are temporary, excavated drainage ways constructed and located to convey runoff to an approved location. The purpose of a temporary swale is to prevent runoff from entering disturbed areas by intercepting and diverting it to a stabilized outlet or to intercept sediment-laden water and divert it to a sediment trapping device. A minimum slope of 0.5% should be maintained. Fig. 5-18 provides a typical detail for a temporary swale. Select size of temporary swale (letter) based on channel stabilization criteria (number) in Table 5-1; slope and drainage are in Table 5-3 (letter-number).

Dike/swales are temporary ridges of soil excavated from an adjoining swale located along the perimeter of the site or disturbed area. The purpose of a perimeter dike/swale is to prevent storm runoff from entering a disturbed area and to prevent sediment-laden storm runoff from leaving the construction site or disturbed area. A perimeter dike/swale can be constructed to divert flows around disturbed areas, along tops of slopes to prevent flows from eroding the slope, or along the base of slopes to direct sediment-laden flows to a sediment trapping device. The perimeter dike/swale should remain in place until the disturbed areas are permanently stabilized. The maximum drainage area for this measure is 0.8 ha (2 ac). Fig. 5-19 provides typical details for a perimeter dike/swale.

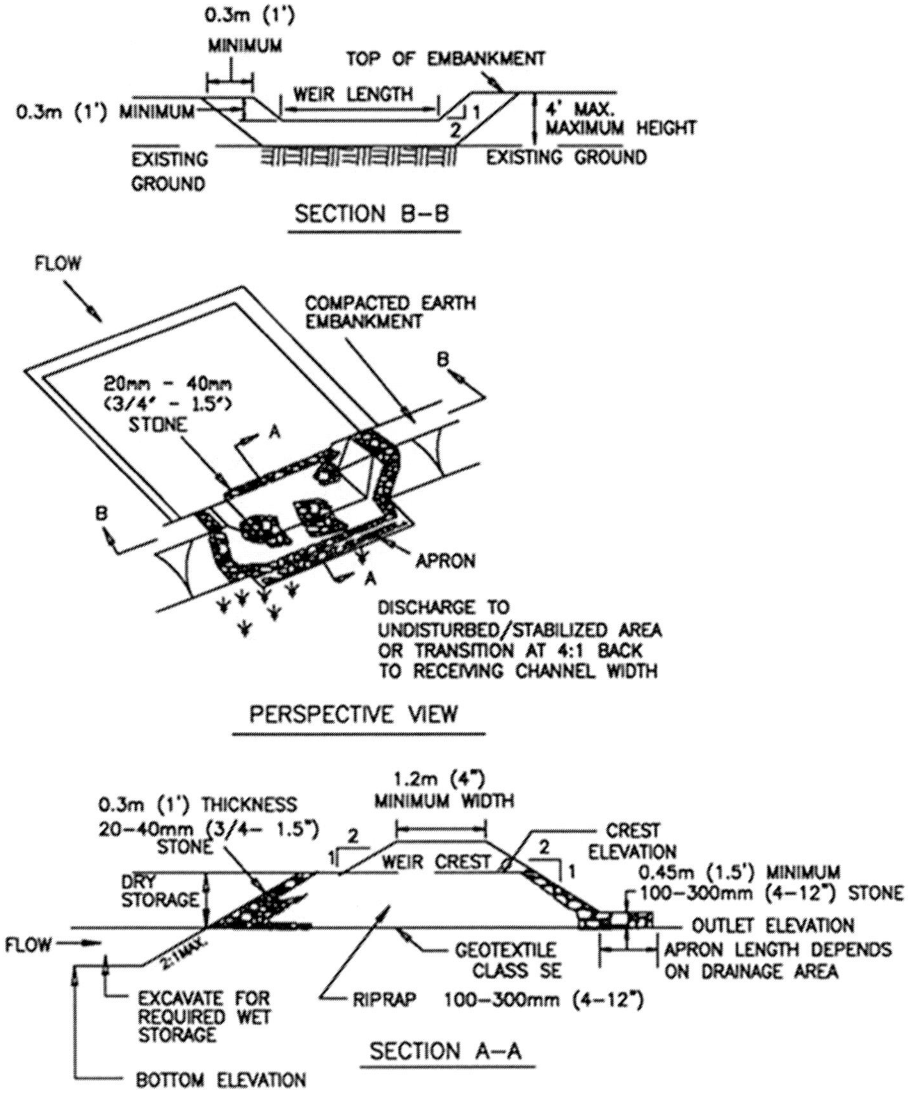

FIGURE 5-15. Riprap/Stone Outlet Sediment Trap
***Source*: Fig. C916, Detail 10A, Riprap/stone outlet sediment trap (ST-IV) (MDE 1994j), with modifications by MPCES Committee, reproduced with permission from the Maryland Department of the Environment.**

Additional construction specifications for stabilization include perimeter dike/swale category 1 (PD/S-1), with up to 0.4 ha (1 ac) seed and mulch. PD/S-2 is categorized as from 0.4 to 0.8 ha (1 to 2 ac) seed and cover with soil stabilization matting or line with sod.

Temporary storm drain diversions are measures used to redirect a storm sewer system or outfall channel to discharge into a sediment trap or basin. The purpose of this measure could be to prevent sediment-laden water conveyed by the storm sewer system from reaching a watercourse or off-site property.

5.4.2 Downdrains and Flumes. Downdrains and flumes are measures used to convey surface runoff on destabilized slopes, and they include (1) the pipe slope drain, and (2) the paved flume (District of Columbia 2003).

A pipe slope drain is a pipe that is installed to convey surface runoff down the face of destabilized slopes. It is used to minimize erosion on the slope face, and use of flexible piping is preferred (MDE 1994n).

Pipe slope drains can be used in conjunction with earth dikes. The dikes direct surface runoff to the slope drain, which conveys concentrated flow down the face of a slope. When used to convey water down a destabilized fill slope on a road construction project the drainage area to the pipe slope drain should be limited to 0.8 ha (2 ac). When used as an inflow protection device, the drainage area should be (no more than) 2 ha (5 ac). Fig. 5-20 provides a typical detail for a pipe slope drain.

Table 5-4 suggests pipe dimensions for various drainage areas. Note that these suggestions do not negate the use of appropriate engineering practices as required by local regulations for sizing of conduits.

MDE (2011) calls for a stone apron of 100 to 180 mm (4 to 7 in.) stone 0.3 m (1 ft) deep over geotextile fabric at the entrance of the pipe but also allows up 2.8 ha (7 ac) of drainage.

A paved flume is a permanent paved channel constructed on a slope. The purpose could be for conveying stormwater runoff safely down the face of a slope without causing erosion problems on or below the slope. This measure typically is used wherever concentrated stormwater runoff should be conveyed from the top

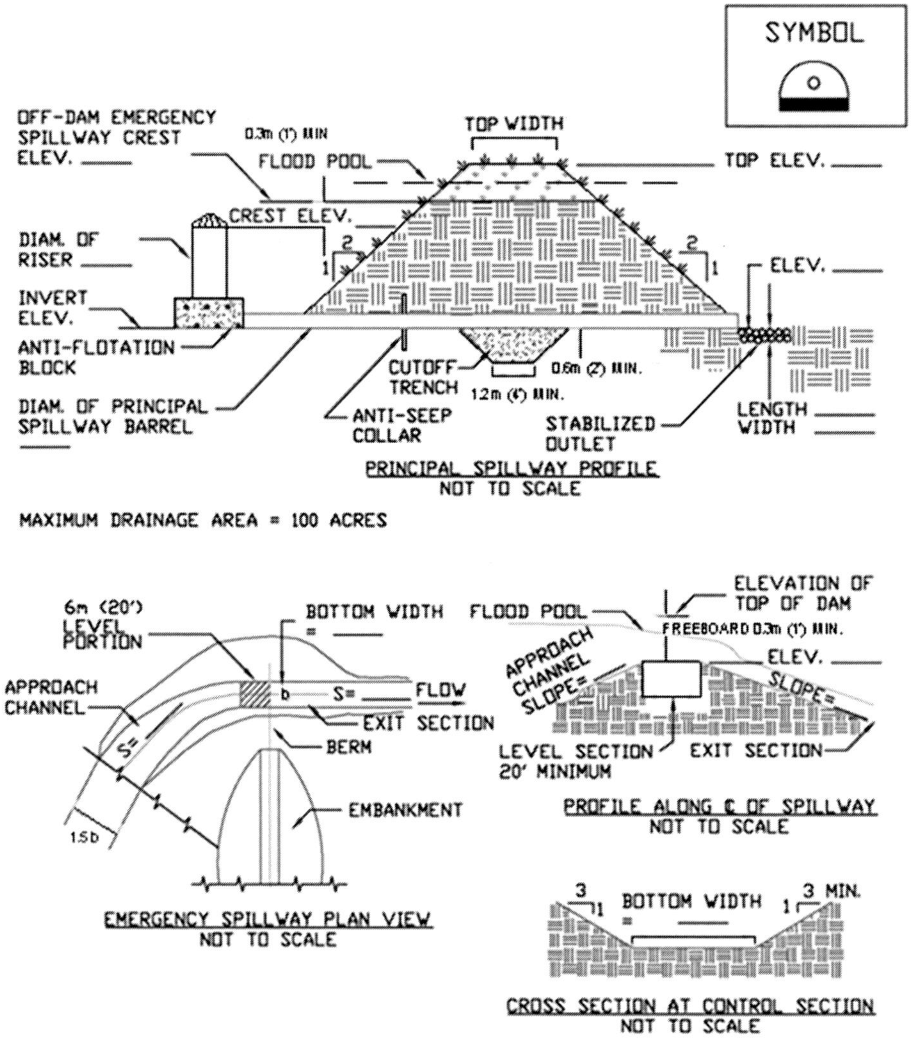

FIGURE 5-16. Sediment Basin Details

***Source*: Fig. 5A.23, Sediment basin detail (NYSDEC 2005), with modifications by MPCES Committee, reproduced with permission from the New York State Department of Environmental Conservation.**

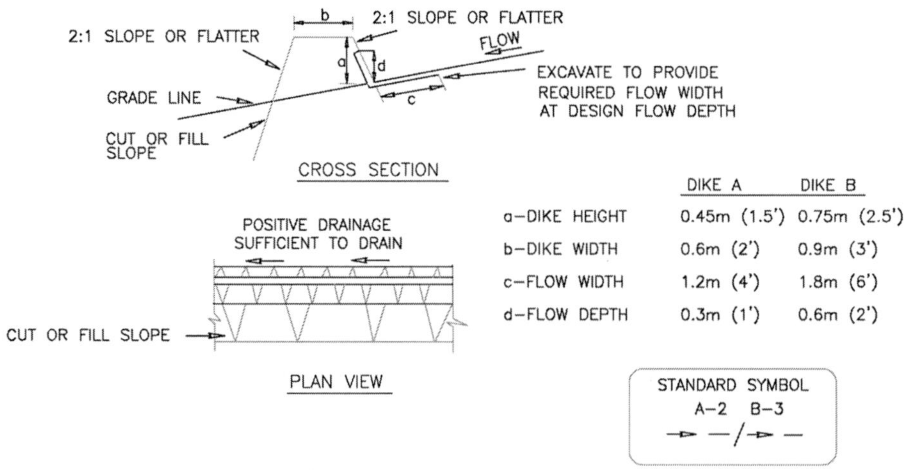

FIGURE 5-17. Earth Dike

***Source*: Fig. A1.6, Detail 1, Earth dike (MDE 1994k), with modifications by MPCES Committee, reproduced with permission from the Maryland Department of the Environment.**

Table 5-1. Stabilization Criteria

Number	Stabilization Measure	Discharge velocity m/s (ft/s)
1	Seed and cover with straw mulch. Not allowed for diverting clear water	
2	Seed and cover with soil stabilization matting or line with sod.	1.2 (4)
3	Use 100–180mm (4–7 in.) stone or recycled concrete equivalent pressed into the soil 180 mm (7 in.) minimum.	1.8 (6)

Source: MDE (1994k); MDE (2011).

Table 5-2. Earth Dike Selection

Slope %	Drainage Area in Hectares [Acres]									
	0.4 (1)	0.8 [2]	1.2 [3]	1.6 [4]	2.0 [5]	2.4 [6]	2.8 [7]	3.2 [8]	3.6 [9]	4.0 [10]
1	A-1	A-1	A-2	A-2	B-2	B-2	B-2	B-2	B-2	B-2
2	A-1	A-2	A-2	A-2	A-2	B-2	B-2	B-2	B-2	B-2
3	A-1	A-2	A-2	A-2	A-2	A-2	B-2	B-3	B-3	B-3
4	A-2	A-2	A-2	A-2	A-2	A-2	A-3	B-3	B-3	B-3
5	A-2	A-2	A-2	A-2	A-3	A-3	A-3	B-3	B-3	B-3
6	A-2	A-2	A-2	A-3	A-3	A-3	A-3	B-3	B-3	B-3
7	A-2	A-2	A-3	A-3	A-3	A-3	A-3	B-3	B-3	B-3
8	A-2	A-2	A-3	A-3	A-3	A-3	A-3	A-3	NA	NA
9	A-2	A-2	A-3	A-3	A-3	A-3	A-3	A-3	NA	NA
10	A-2	A-3	A-3	A-3	A-3	A-3	A-3	A-3	NA	NA

Source: MDE (1994k).

to the bottom of cut or fill slopes on a permanent basis, and a riprap-lined channel is not capable of conveying the runoff without erosion.

Paved flumes have been used routinely on highway cuts and fills to convey concentrated stormwater runoff from the top to the bottom of the slope without erosion. Standards and specifications have been developed for these structures, which apply to all secondary and primary highway construction projects. Fortunately, these structures have equal applicability to cut-and-fill slopes for construction projects other than highways. Fig. 5-21 provides a typical detail for a paved flume. The maximum slope for this measure is 1.5/1 (VDEQ 1992f).

Table 5.5 provides specifications for straight, #4 steel reinforcing bars.

The discharge of a paved flume should be protected from erosion by previously described measures; however, the use of an energy dissipater to reduce velocity of flow is highly recommended to prevent undermining of the flume and provide a stable transition zone between flume and ultimate discharge point (VDEQ 1992f).

5.5 GRASS OR LINED WATERWAYS

This measure consists of a permanent, designed waterway, shaped, sized, and lined with appropriate vegetation or structural material, used to safely convey stormwater runoff within or away from a developing area. The purpose is to provide for the conveyance of concentrated surface runoff water to a receiving channel or system without damage from erosion.

This measure is generally applicable to artificially constructed channels, including roadside ditches and intermittent natural channels, which are constructed or are modified to accommodate

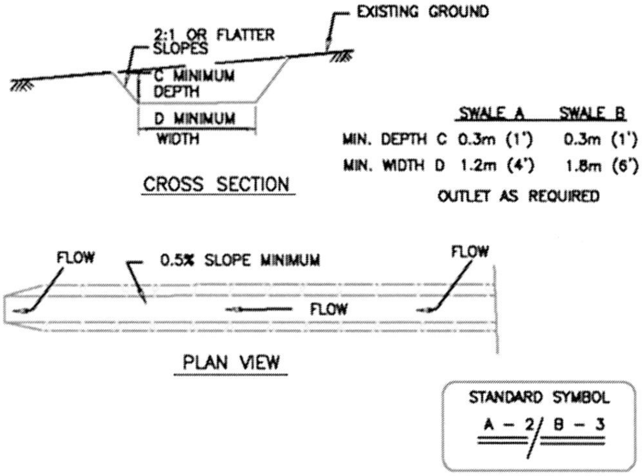

FIGURE 5-18. Temporary Swale
***Source*: Fig. A24, Detail 2, Temporary swale (MDE 1994l), with modifications by MPCES Committee, reproduced with permission from the Maryland Department of the Environment.**

Table 5-3. Temporary Swale Selection

Slope %	Drainage Area in Hectares [Acres]									
	0.4 (1)	0.8 [2]	1.2 [3]	1.6 [4]	2.0 [5]	2.4 [6]	2.8 [7]	3.2 [8]	3.6 [9]	4.0 [10]
1	A-1	A-1	A-1	A-1	A-1	B-2	B-2	B-2	B-2	B-2
2	A-1	A-1	A-2	A-2	A-2	B-2	B-2	B-2	B-2	B-2
3	A-1	A-2	A-2	A-2	A-2	B-2	B-2	B-2	B-3	B-3
4	A-1	A-2	A-2	A-2	A-2	B-2	B-2	B-3	B-3	B-3
5	A-2	A-2	A-2	A-2	A-2	B-2	B-3	B-3	B-3	B-3
6	A-2	A-2	A-2	A-2	A-2	B-3	B-3	B-3	B-3	B-3
7	A-2	A-2	A-2	A-2	A-3	B-3	B-3	B-3	B-3	B-3
8	A-2	A-2	A-3	A-3	A-3	B-3	B-3	B-3	B-3	B-3
9	A-2	A-2	A-3	A-3	A-3	B-3	B-3	B-3	B-3	B-3
10	A-2	A-2	A-3	A-3	A-3	B-3	B-3	B-3	B-3	B-3

Source: MDE (2011), reproduced with permission from the Maryland Department of the Environment.

flows generated by land development. The implementation of this control should come only after a channel adequacy analysis for capacity and velocity has been performed. The measure should be installed and stabilized prior to the introduction of postdevelopment flows. This measure generally is not applicable to continuous flowing natural streams, as major streams require full design considerations and calculations and are governed by other regulations. Fig. 5-22 provides details for typical cross sections of waterways.

5.6 INLET AND OUTLET PROTECTION

Inlet and outlet protection practices are a group of control practices developed to provide protection at critical inflow and outlet points of stormwater conveyance practices. This group of practices includes riprap inflow protection, gabion inflow protection, stone check dams, stone outlet protection, and rock outlet protection (District of Columbia 2003).

5.6.1 Riprap Inflow Protection. A riprap inflow protection practice is a temporary or permanent lined drainage way installed to convey concentrated runoff into sediment traps and basins or down steep slopes as applicable. Riprap inflow protection

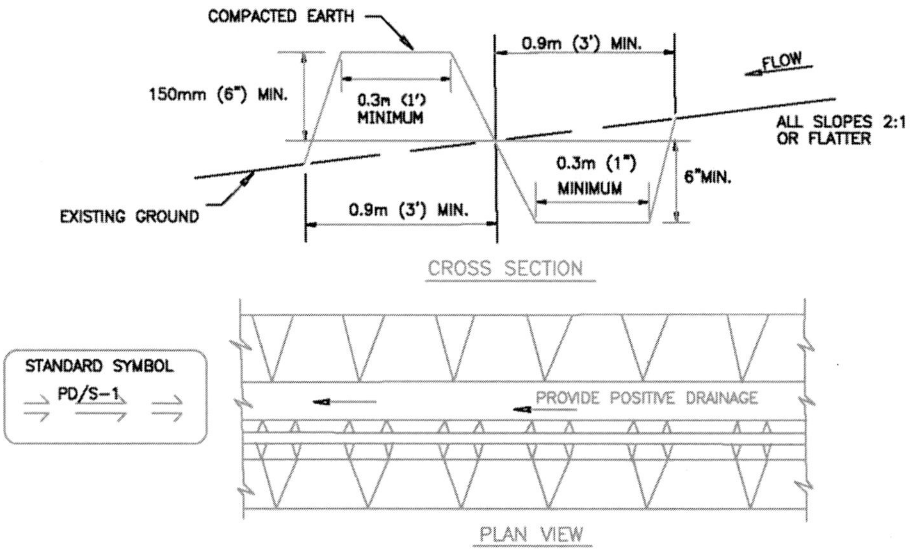

FIGURE 5-19. Perimeter Dike/Swale
***Source*: Fig. A33, Detail 3, Perimeter dike/swale (MDE 1994m), with modifications by MPCES Committee, reproduced with permission from the Maryland Department of the Environment.**

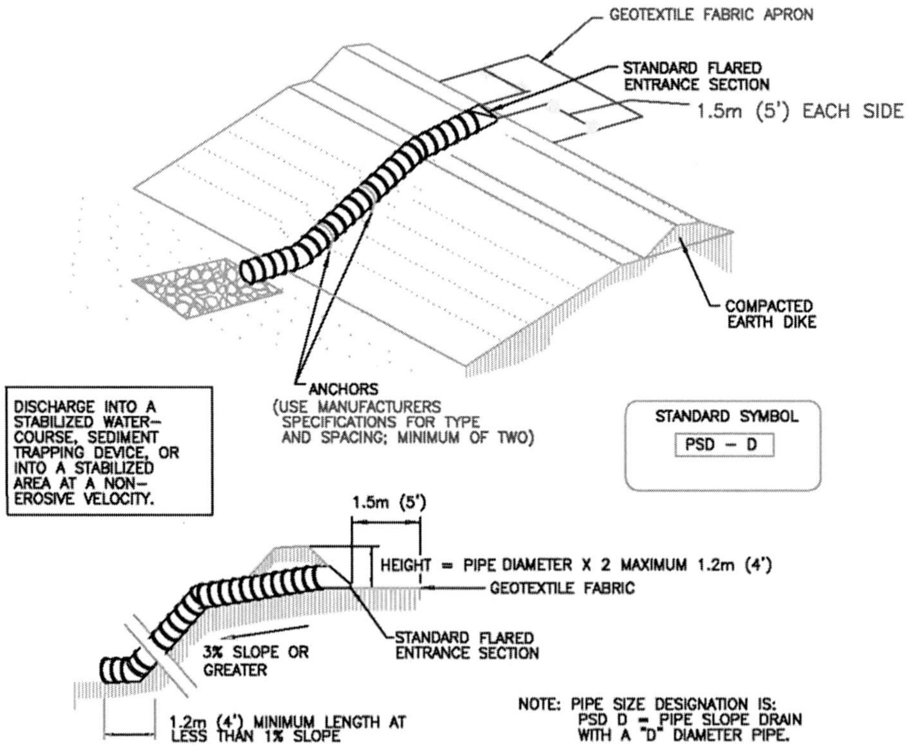

FIGURE 5-20. Pipe Slope Drain
***Source*: Fig. B54, Detail 4, Pipe Slope Drain (MDE 1994n), with modifications by MPCES Committee, reproduced with permission from the Maryland Department of the Environment.**

consists of the installation of rock or recycled concrete equivalent in a flow channel for stabilization. The purpose of riprap inflow protection is to provide stable conveyance of concentrated runoff down steep slopes (i.e., into temporary sediment traps and basins), thereby preventing erosion of the flow channel. Fig. 5-23 provides a typical detail for a riprap inflow protection practice.

Riprap inflow protection is required where the slope of a drainage way contributing to a sediment trap or basin exceeds 10:1 but is less than 4:1. Runoff may be directed to the inflow

Table 5-4. Pipe Size for Drainage Area of Pipe Slope Drain

Pipe/tubing diameter		Maximum drainage area		Symbol	
Millimeters	Inches	Hectares	Acres	Metric	Customary
300	12	0.2	0.5	PSD-300	PSD-12
460	18	0.6	1.5	PSD-460	PSD-18
530	21	1.0	2.5	PAD-530	PSD-21
610	24	1.4	3.5	PASD-610	PSD-24
610*	24*	2.0	5.0	PSD-610 (2)	PSD-24 (2)

*Note: For drainage areas over 1.4 ha (3−1/2 ac) two 610 mm (24 in.) pipes shall be used; a minimum spacing of 2D 1.3 m (4 ft) is required between pipes.
Source: MDE (1994o), reproduced with permission from Maryland Department of the Environment.

Table 5-5. Specifications for Placement of Reinforcing Steel

Mark	Number*	Length in mm [in.] Slope 2:1	Length in mm [in.] Slope 1.5:1	Spacing C-C mm [in.]
Bars F	3	360 [14]	360 [14]	100 [4]
Bars G	9	430 [17]	430 [17]	150 [6]
Dowels	10	460 [18]	460 [18]	125 [5]

*Note: Numbers shown are for one anchor lug, curtain wall and joint.
Source: VDEQ (1992f).

device by means of dikes or swales. Riprap-lined inflow channels should be 0.3 m (1 ft) in depth, have a trapezoidal cross section with 2:1 or flatter side slopes and minimum 1.2 m (4 ft) bottom width. Blend riprap into the ground.

5.6.2 Gabion Inflow Protection. Gabion inflow protection is a temporary or permanent lined drainage way installed to convey concentrated runoff into sediment traps and basins or down steep slopes as applicable. It consists of the installation of 300 mm (12 in.) thick gabions filled with 100 to 180 mm (4 to 7 in.) stone placed in the flow channel for stabilization. The purpose of gabion inflow protection is to provide stable conveyance of concentrated runoff down steep slopes, from 4:1 to 2:1 (maximum) (MDE 2011), into temporary sediment traps and basins, thus preventing erosion of the flow channel.

Surface runoff may be directed to the inflow device by means of dikes or swales. Suggested limitation on drainage area is

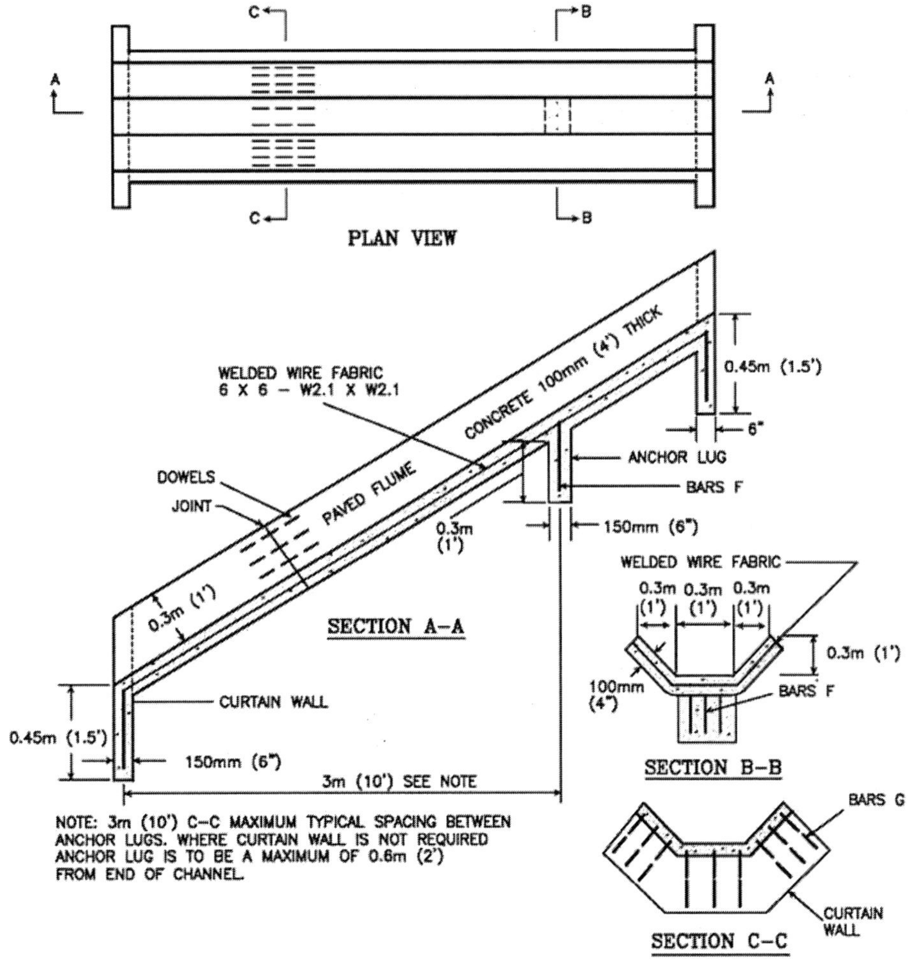

FIGURE 5-21. Paved Flume
Source: Plate 3.16-1, Paved flume (VDEQ 1992f), with modifications by MPCES Committee, reproduced with permission from the Virginia Department of Environmental Quality.

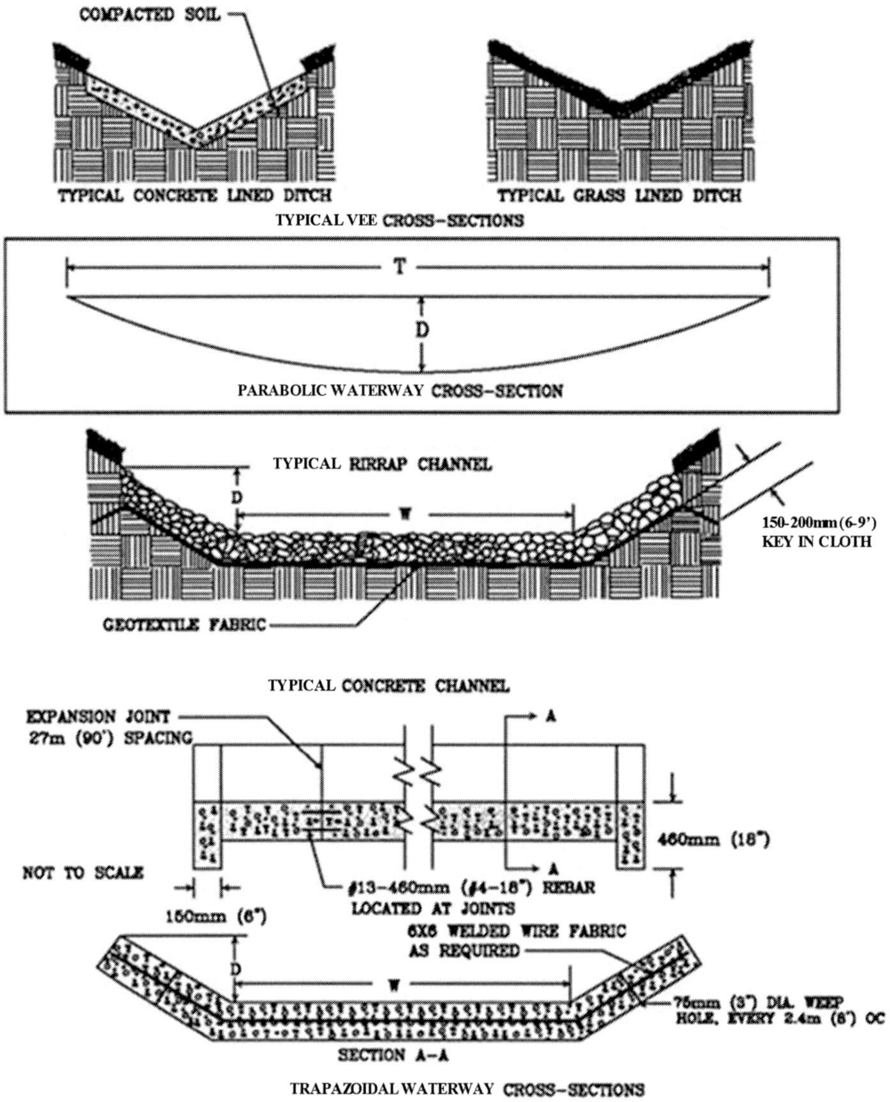

FIGURE 5-22. Typical Waterway

***Source*: Plate 3.17-1, Typical waterway cross sections (VDEQ 1992g), with modifications by MPCES Committee, reproduced with permission from the Virginia Department of Environmental Quality.**

4 ha (10 ac) (MDE 2011). Fig. 5-24 provides a typical detail for a gabion inflow protection practice.

5.6.3 Stone Check Dams. Stone check dams are stone weirs placed in series in swales or ditches. Stone check dams are constructed to reduce runoff velocities to nonerosive rates and to prevent channel erosion in drainage courses. Fig. 5-25 provides a typical detail for a stone check dam.

5.6.4 Stone Outlet Structures. A stone outlet structure is a temporary stone dike installed in conjunction with and as a part of an earth dike. The purpose of the temporary stone outlet structure is to filter sediment-laden runoff, provide a protected outlet for an earth dike, provide for diffusion of concentrated flow, and allow the area behind the dike to dewater and filter sediment-laden runoff. Fig. 5-26 provides a typical detail of a stone outlet structure.

Stone outlet structures apply to any point of discharge where there is need to dispose of runoff at a protected outlet or to diffuse concentrated flow for the duration of the period of construction.

5.6.5 Rock Outlet Protection. Rock outlet structures consist of rocks placed at the outfall of channels or culverts to reduce the velocity of flow in the receiving channel to nonerosive rates. This practice applies where discharge velocities and energies at the outlets of culverts are sufficient to erode the next downstream reach. This applies to outlets of all types, such as sediment basins, stormwater management ponds, and road culverts. Fig. 5-27 provides typical details of a rock outlet structure.

When there is no end wall, construct the upstream end of the apron so that the width is twice the diameter of the outlet pipe, extending the stone under the outlet by at least 0.45 m (1.5 ft) (MDE 2011).

5.7 TEMPORARY IN-STREAM CONSTRUCTION MEASURES

Construction activities often require crossing of small streams and waterways. These are very sensitive environmental site features, and a number of ESC practices have been developed

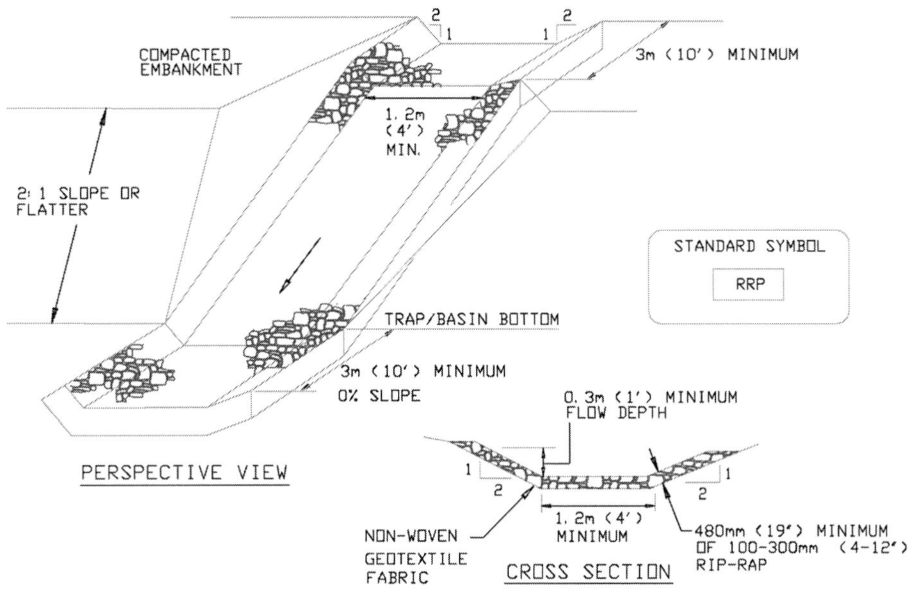

FIGURE 5-23. Riprap Inflow Protection

***Source*: Fig. B-62, Detail 5, Riprap inflow protection (MDE 1994o), with modifications by MPCES Committee, reproduced with permission from the Maryland Department of the Environment.**

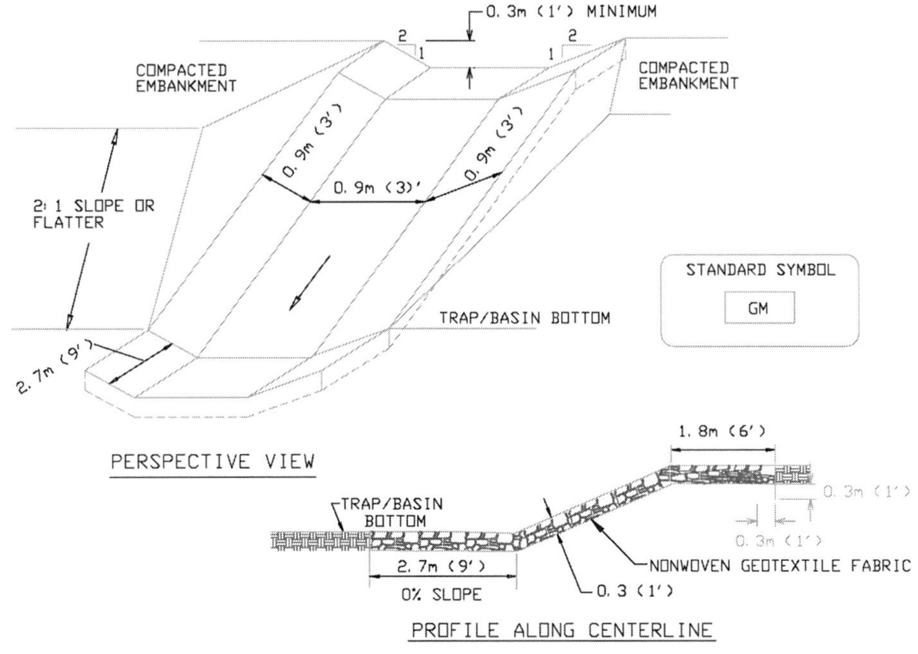

FIGURE 5-24. Gabion Inflow Protection

***Source*: Fig. B-72, Detail 6, Riprap inflow protection (MDE 1994p), with modifications by MPCES Committee, some of which reflect current MDE guidance (2011), reproduced with permission from the Maryland Department of the Environment.**

to control erosion and sediment in this type of construction. The available measures include pump around practice, culvert pipe with access road, diversion pipe, sandbag/stone channel diversion, fabric-based channel diversion, and turbidity curtain (District of Columbia 2003).

5.7.1 Pump Around Practice. A pump around practice is a temporary measure for dewatering in-channel construction sites. The practice consists of installing a temporary pump around and supporting measures to divert flow around in-stream construction sites. Fig. 5-28 illustrates a typical pump around installation.

5.7.2 Culvert Pipe with Access Road. A culvert pipe with access road is a temporary measure for providing access to stream enhancement sites. The practice consists of installing a culvert pipe and associated access road for the purpose of

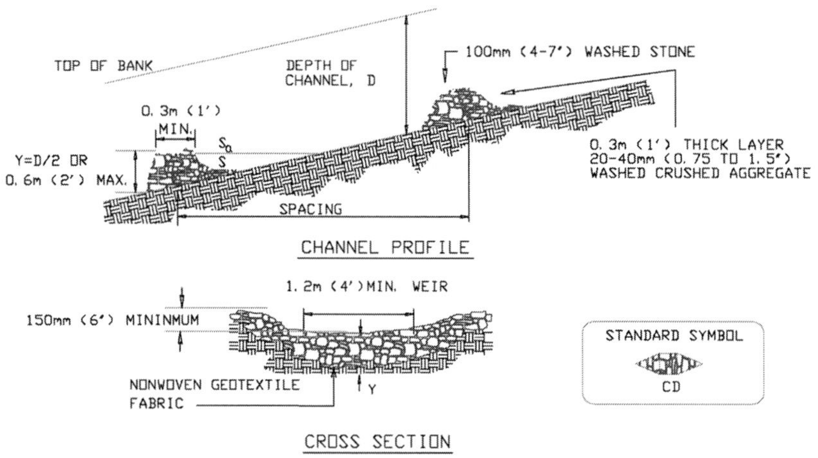

FIGURE 5-25. Stone Check Dam

***Source*: Fig. B-83, Detail 7, Stone check dam (MDE 1994q), with modifications by MPCES Committee, reproduced with permission from the Maryland Department of the Environment.**

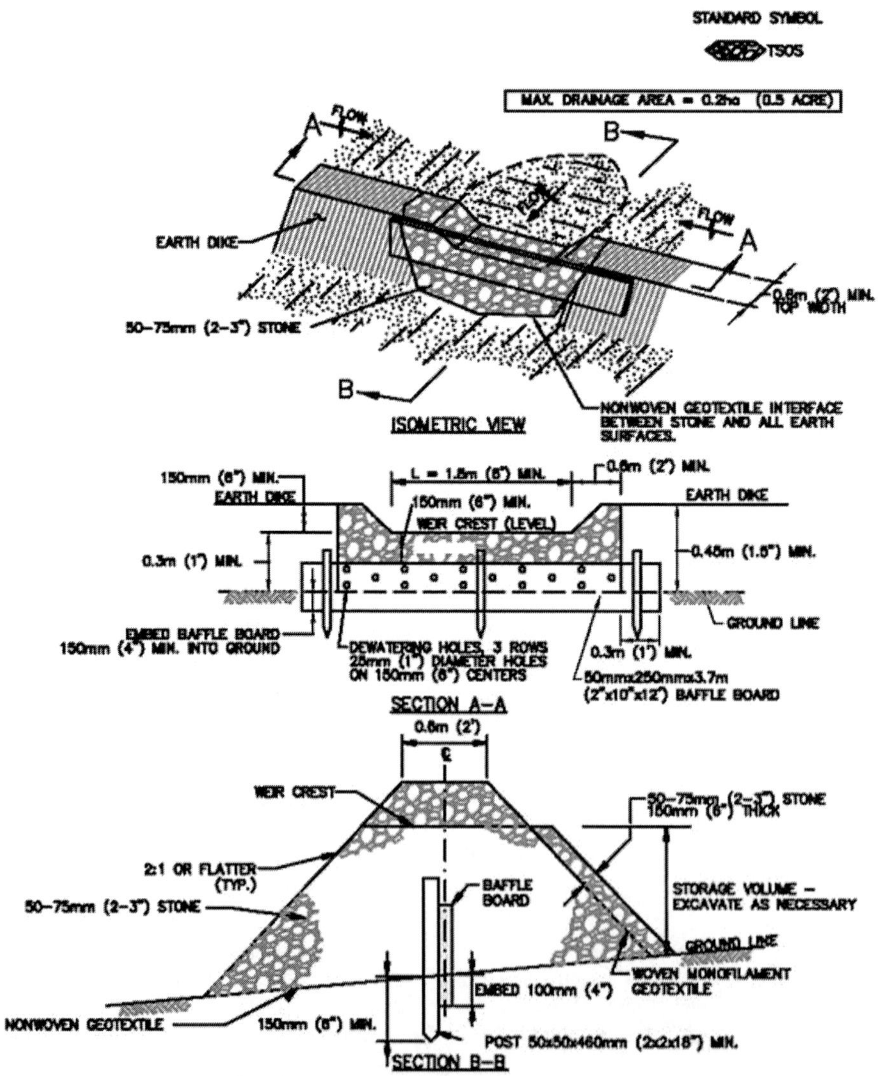

FIGURE 5-26. Temporary Stone Outlet Structures

***Source*: Detail E-7, Stone check dam (MDE 2011), with modifications by MPCES Committee, reproduced with permission from the Maryland Department of the Environment.**

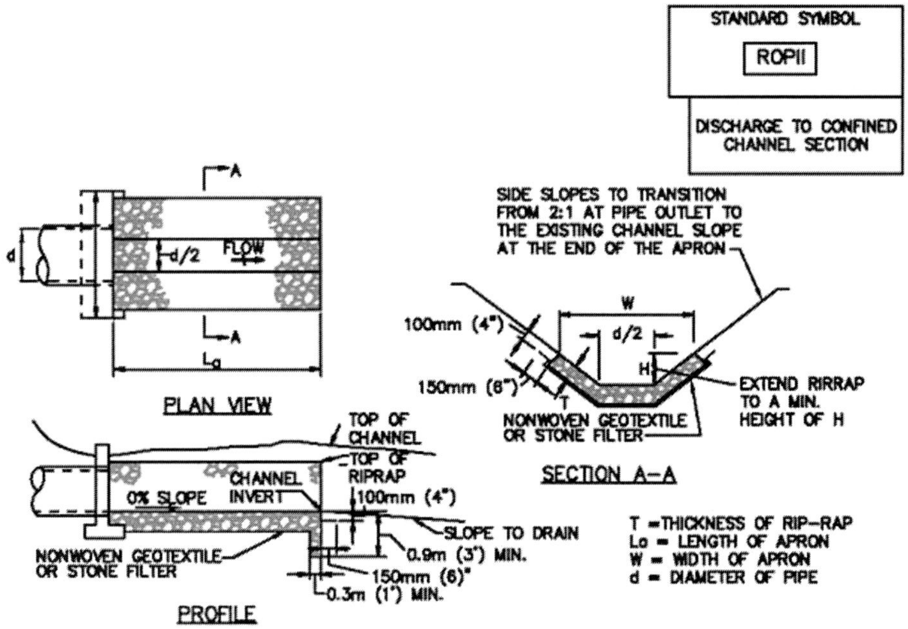

FIGURE 5-27. Rock Outlet Protection

***Source*: Detail D-4-1-B, Rock outlet protection II (MDE 2011), with modifications by MPCES Committee, reproduced with permission from the Maryland Department of the Environment.**

erosion control when construction activities occur within the stream corridor. Figure 5-29 provides a typical detail for this practice.

Culvert pipes with access roads can be used effectively for installation of utility lines at stream crossings. Diversions that have an insufficient flow capacity can fail and severely erode the disturbed channel section under construction. Therefore, in-channel construction activities should occur only during periods of low rainfall.

5.7.3 Diversion Pipe. A diversion pipe is a temporary measure for dewatering in-channel construction sites. The practice consists of installing flow diversion pipes in combination with sandbag or stone diversions when construction activities occur within the stream channel. Figure 5-30 provides a typical detail for a diversion pipe.

Additional construction details include the following (MDE 2011):

- Flexible pipe is preferred but other pipe can be used (make sure joints are watertight).
- Use a double row of sandbags. Materials for sandbags should be resistant to ultraviolet radiation, tearing, and puncture and woven tightly enough to prevent leakage of fill material.
- Impermeable sheeting should be a minimum of 0.254 mm thick (10 mil), resistant to ultraviolet radiation and puncturing and tearing.
- Height of sandbag dike is twice the pipe diameter (D).

Diversion pipes with insufficient flow capacity can cause the channel diversion to fail, resulting in severe erosion of the disturbed channel section under construction. Therefore, in-channel construction activities should occur only during periods of low flow.

5.7.4 Channel Diversion. A channel diversion is a temporary measure for dewatering in-channel construction sites. The practice consists of installing sandbag or stone flow diversions for the purpose of erosion control when construction activities occur within the stream channel. Figure 5-31 provides a typical detail for a channel diversion.

Diversions are used to isolate work areas from flow during the construction of in-stream projects. Diversions that have an insufficient flow capacity can fail and severely erode the disturbed channel section under construction. Therefore, in-channel construction activities should occur only during periods of low rainfall. This temporary measure may not be practical in large channels.

5.7.5 Fabric-Based Channel Diversion. A fabric-based channel diversion is a temporary measure for dewatering in-channel construction sites. The practice consists of installing fabric-based diversion channels for the purpose of erosion control when construction activities occur within the stream channel. Figure 5-32 provides a typical detail for a fabric-based channel diversion.

Diversions are used to divert flow during construction of in-stream projects. Diversions that have an insufficient flow capacity can fail and severely erode the disturbed channel section under construction. Therefore, in-channel construction activities should occur only during periods of low rainfall.

5.7.6 Turbidity Curtain. A turbidity curtain is a temporary measure made of floating geotextile material that minimizes sediment transport from a disturbed area adjacent to or within a body of water. The purpose of this practice is to provide sediment protection for a watercourse from up-slope land disturbance or from dredging or filling within the watercourse.

5.8 DEWATERING PRACTICES

Dewatering refers to the act of removing and discharging water from excavated areas on construction sites or from sediment traps or basins on construction sites. A number of standard practices have been developed to achieve this function. They include

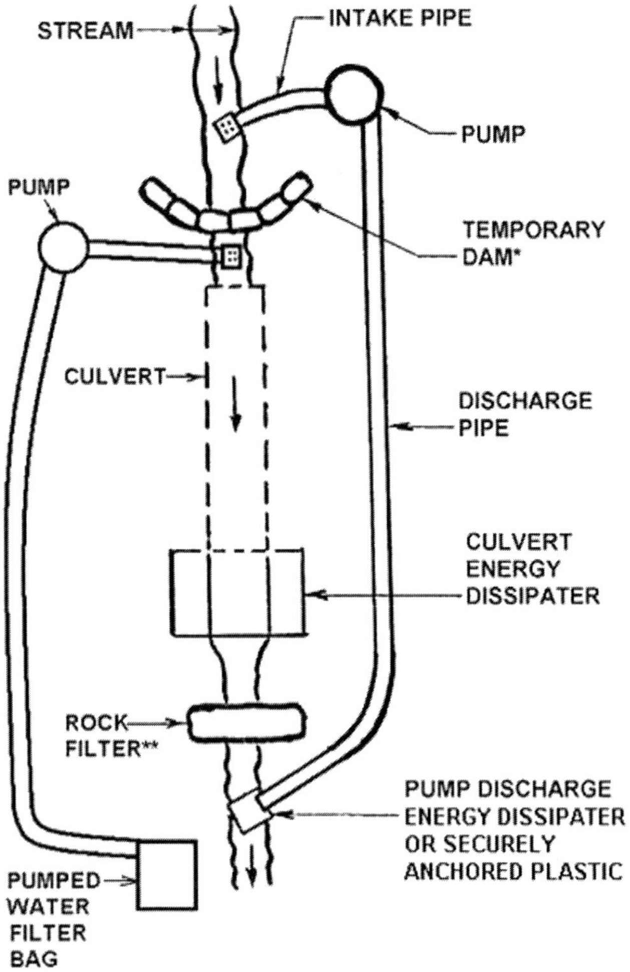

FIGURE 5-28. Pump Around Practice
***Source*: Fig. 3.11. Temporary cofferdam and pump bypass around in-channel work areas (PADEP 2012), with modifications by MPCES Committee, reproduced with permission from the Pennsylvania Department of Environmental Protection, http://www.elibrary.dep.state.pa.us/dsweb/Get/Document-88925/363-2134-008.pdf.**
Note: *Sandbags, Jersey barriers or other nonerosive material, no earth fill. Do not excavate a sump for the pump intake.
****For low-gradient channels, the rock filter may be replaced by an impervious cofferdam to prevent backflow into the work area.**

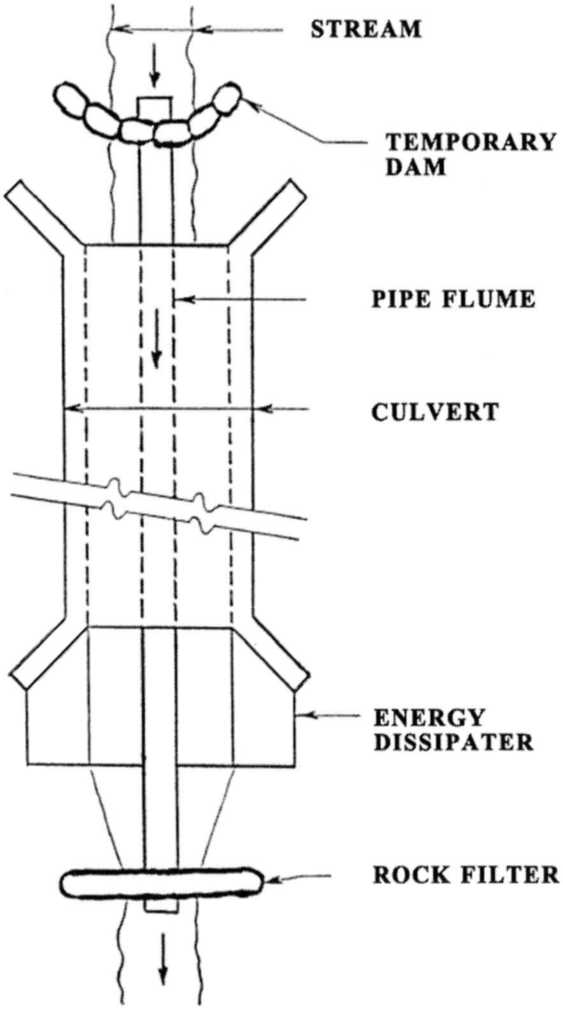

FIGURE 5-29. Culvert Pipe with Access Road
***Source*: Fig. 3.10. Rigid or flexible pipe flume through a work area, (PADEP 2012), with modifications by MPCES Committee, reproduced with permission from the Pennsylvania Department of Environmental Protection, http://www.elibrary.dep.state.pa.us/dsweb/Get/Document-88925/363-2134-008.pdf.**

removable pumping station, sump pit, pumped water filter bag, sediment tanks, and dewatering basins (District of Columbia 2003).

These standard practices apply to removal and discharge of water from any excavated area or sediment trap or basin at any construction site. Given the unique conditions at any particular construction site, any or all of the practices may apply. Regardless of the applicability of the practices listed herein, operators should be required to use acceptable procedures for maintenance and dewatering, as well as local regulations and guidance. In all cases, every effort should be made to eliminate sediment pollution associated with dewatering.

Designers should specify the preferred procedures for dewatering on their plans. In particular, designers should identify procedures for dewatering sediment traps and basins prior to elimination of the last sediment control facility on the site or prior to conversion of sediment control facilities to stormwater management facilities. Recommended procedures should be consistent with these standards. Atypical site conditions may require innovative dewatering designs.

5.8.1 Removable Pumping Station. A removable pumping station is a temporary structure that is used to remove water from excavated areas, sediment traps, and basins. The removable pumping station filters sediment-laden water at a pump intake, prior to discharging to a suitable area. The station requires frequent maintenance; pull out the inner pipe and replace the geotextile when clogged (MDE 2011). Fig. 5-33 provides a typical detail for a removable pumping station.

Removable pumping stations are constructed where water collects and should be pumped away during excavation, cofferdam dewatering, maintenance or removal of sediment traps and basins, or for other uses as applicable. These are preferred over sump pits on projects where a long duration of pumping is expected.

5.8.2 Sump Pit. A sump pit is a temporary pit from which excess water is pumped. Sump pits are constructed when water collects and should be pumped away during excavating, cofferdam dewatering, maintenance or removal of sediment traps and basins, or other uses as applicable. The sump pit filters water

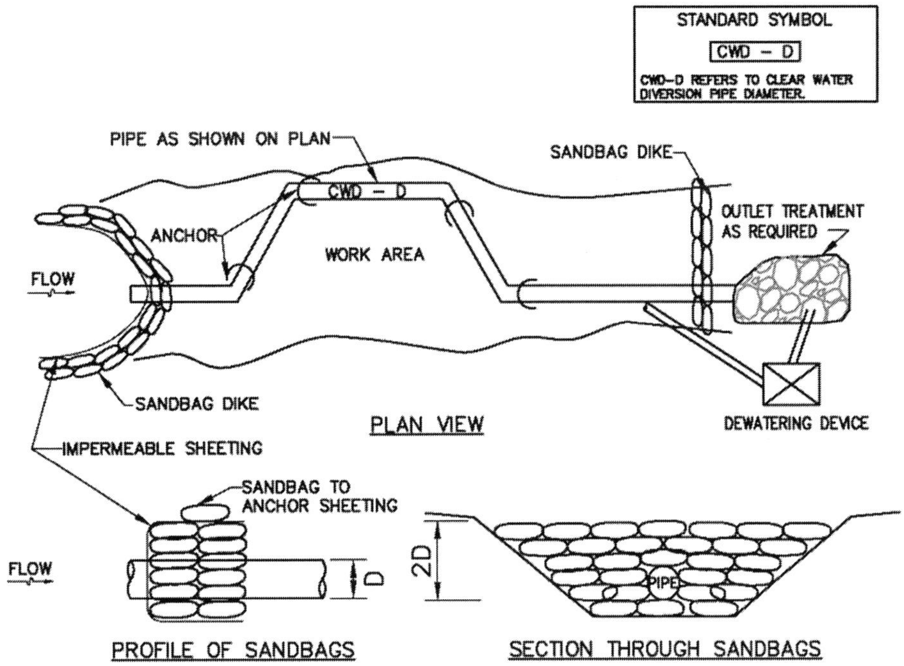

FIGURE 5-30. Diversion Pipe
***Source*: Detail C-6, Clear water diversion pipe (MDE 2011), with modifications by MPCES Committee, reproduced with permission from the Maryland Department of the Environment.**

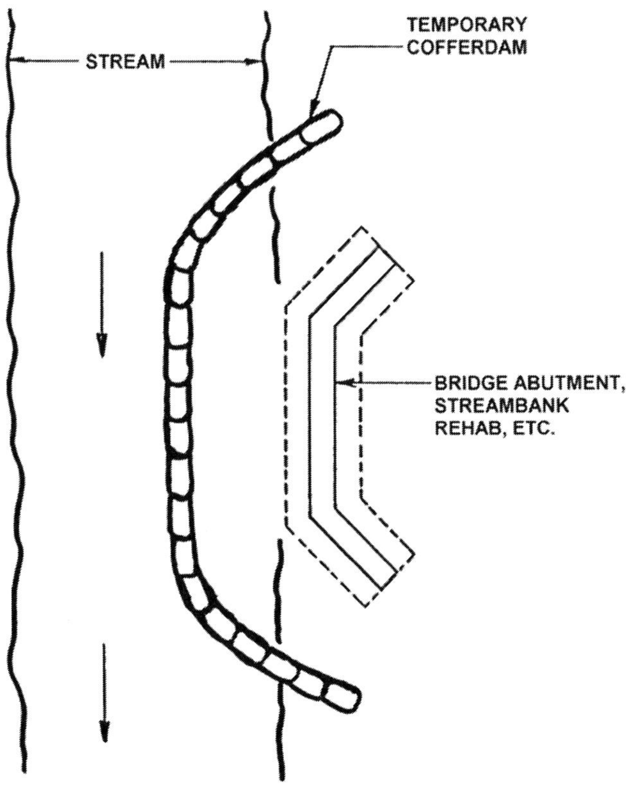

FIGURE 5-31. Channel Diversion
***Source*: Fig. 3.12. In-stream cofferdam diversion (PADEP 2012), with modifications by MPCES Committee, reproduced with permission from the Pennsylvania Department of Environmental Protection.**

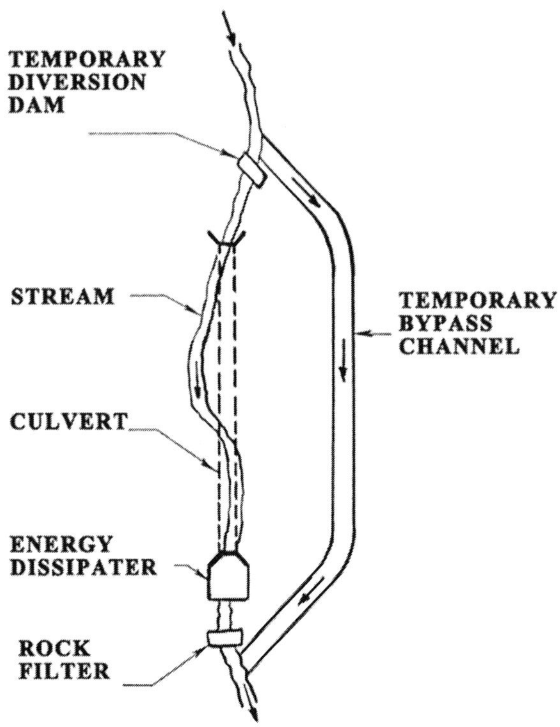

FIGURE 5-32. Fabric-Based Diversion
***Source*: Fig. 3.9. Bypass channel with nonerosive lining (PADEP 2012), with modifications by MPCES Committee, reproduced with permission from the Pennsylvania Department of Environmental Protection.**

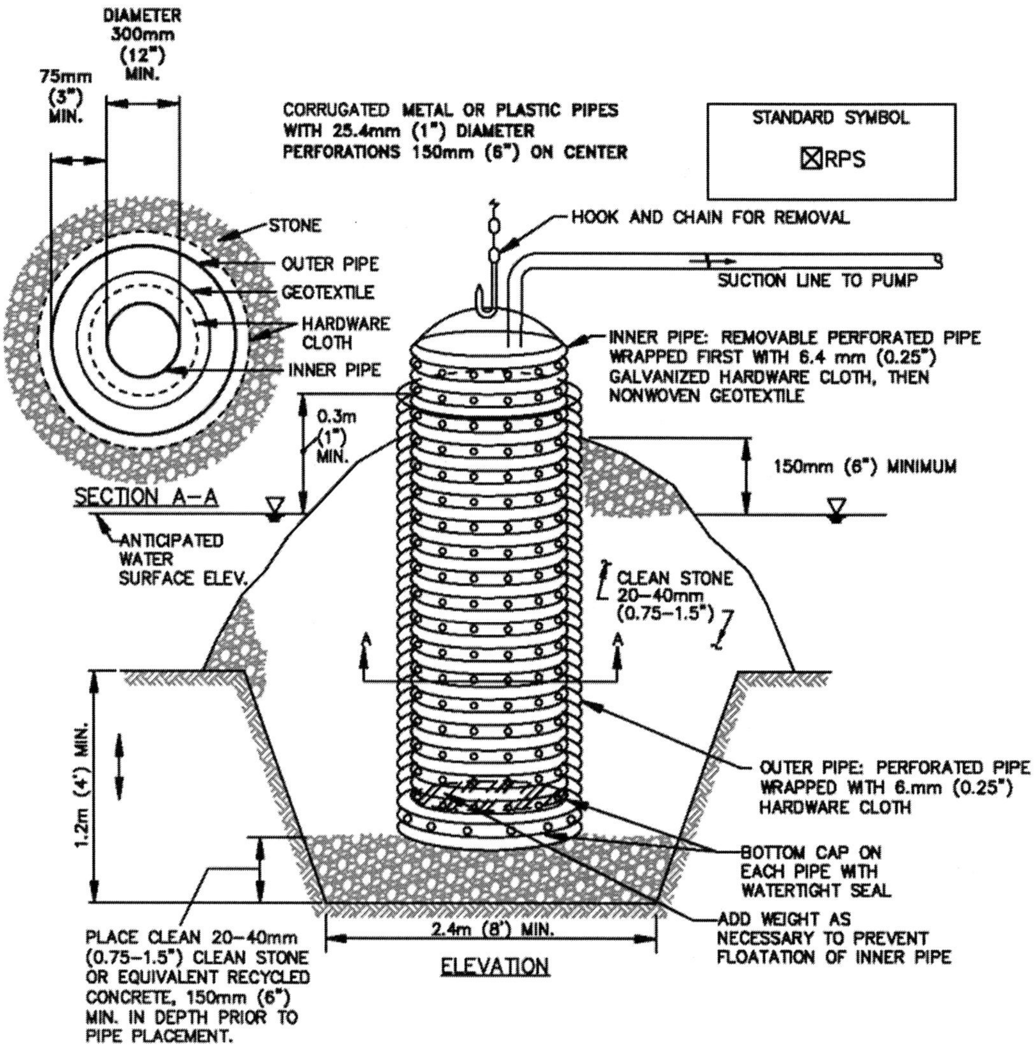

FIGURE 5-33. Removable Pumping Station

Source: **Detail F-1, Removable pumping station (MDE 2011), with modifications by MPCES Committee, reproduced with permission from the Maryland Department of the Environment.**

being pumped to reduce sedimentation to receiving streams. A sump pit can require frequent maintenance and if clogged, remove perforated pipe and replace geotextile, and even stone, if required (MDE 2011). Fig. 5-34 provides a typical detail for a sump pit.

5.8.3 Pumped Water Filter Bag. A pumped water filter bag is a temporary filtering device for water that is discharged from dewatering activities and used to filter sediment-laden water prior to the water being discharged off site. Fig. 5-35 provides typical details for a pumped water filter bag.

Filter bags may be used to filter water pumped from disturbed areas prior to discharging to a receiving water body. They may also be used to filter water pumped from the sediment storage areas of sediment basins: (1) Control pumping rate to prevent excessive pressure within the filter bag in accordance with the manufacturer recommendations. As the bag fills with sediment, reduce pumping rate, (2) Remove and properly dispose of filter bag at completion of pumping operations or after bag has reached capacity, whichever occurs first, (3) Replace filter bag if bag clogs or has rips, tears, or punctures. During operation keep connection between pump hose and filter bag water tight. Replace bedding if it becomes displaced.

5.8.4 Sediment Tank. A sediment tank is a compartmented tank container through which sediment-laden water is pumped to trap and retain the sediment prior to pumping the water to drainage ways, adjoining properties, and rights-of-way below the sediment tank site. Fig. 5-36 provides a typical detail for a sediment tank.

A sediment tank is to be used on sites where excavations are deep and space is limited, such as urban construction, where direct discharge of sediment-laden water to stream and storm drainage systems is to be avoided.

5.8.5 Dewatering Basin. A dewatering basin is a temporary measure for settling of sediment-laden water. The practice consists of installing dewatering basins jointly with channel diversion measures to settle sediment from the laden water from in-stream construction sites before the water re-enters the downstream reach. This practice also may be used for work outside of streams. Fig. 5-37 provides typical details for a dewatering basin.

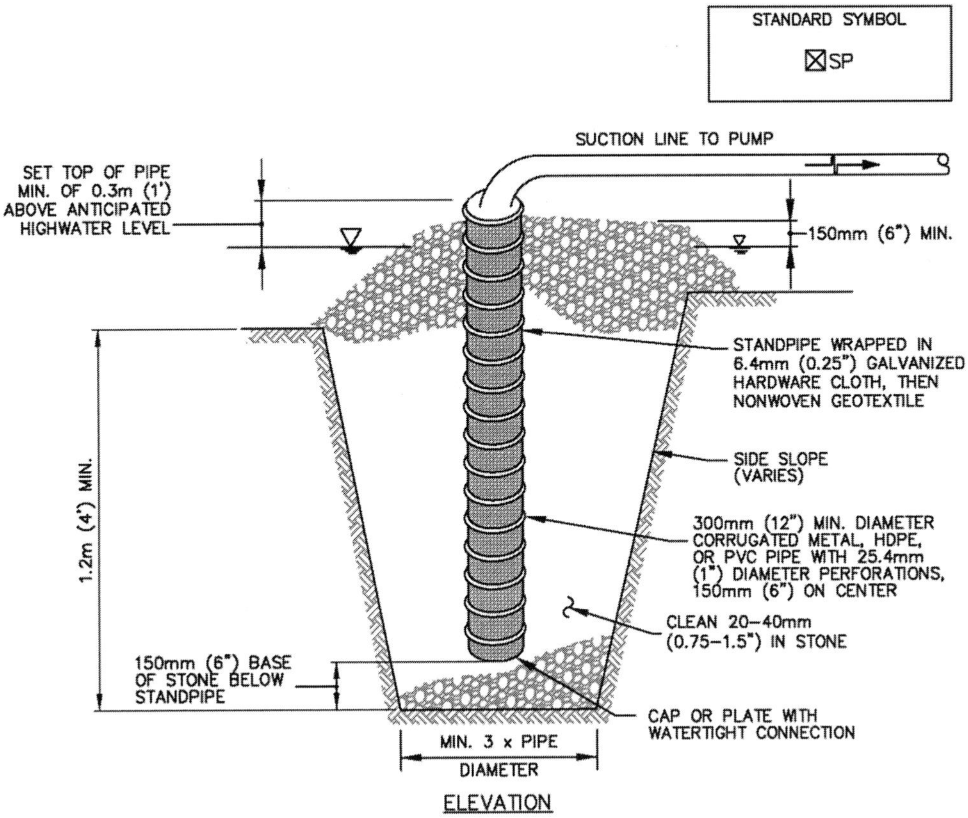

FIGURE 5-34. Sump Pit
***Source*: Detail F-2, Sump pit (MDE 2011), with modifications by MPCES Committee., reproduced with permission from the Maryland Department of the Environment.**

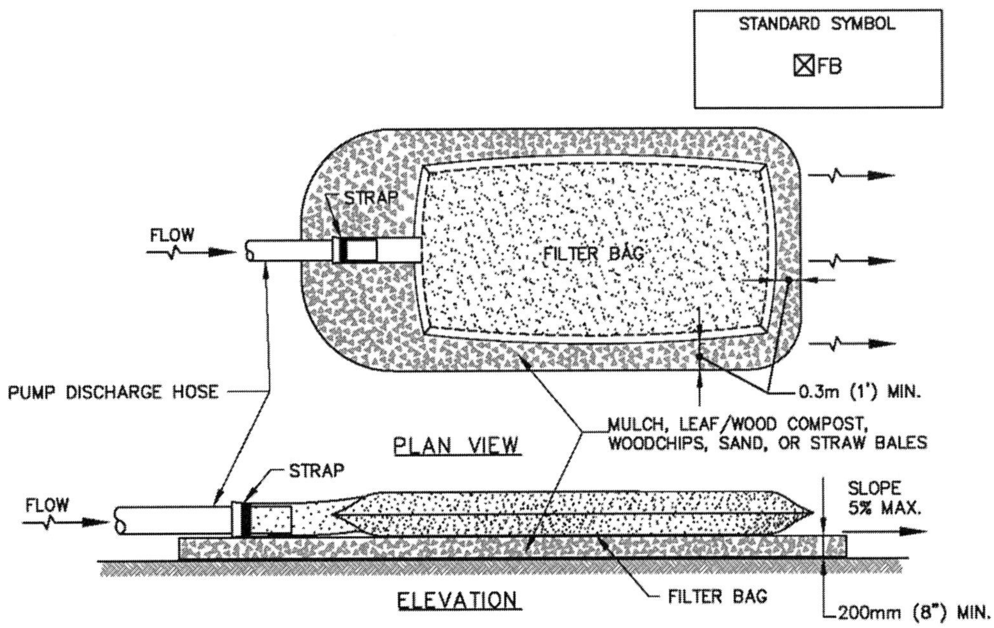

FIGURE 5-35. Pumped Water Filter Bag
***Source*: Detail F-4, Filter bag (MDE 2011), with modifications by MPCES Committee, reproduced with permission from the Maryland Department of the Environment.**

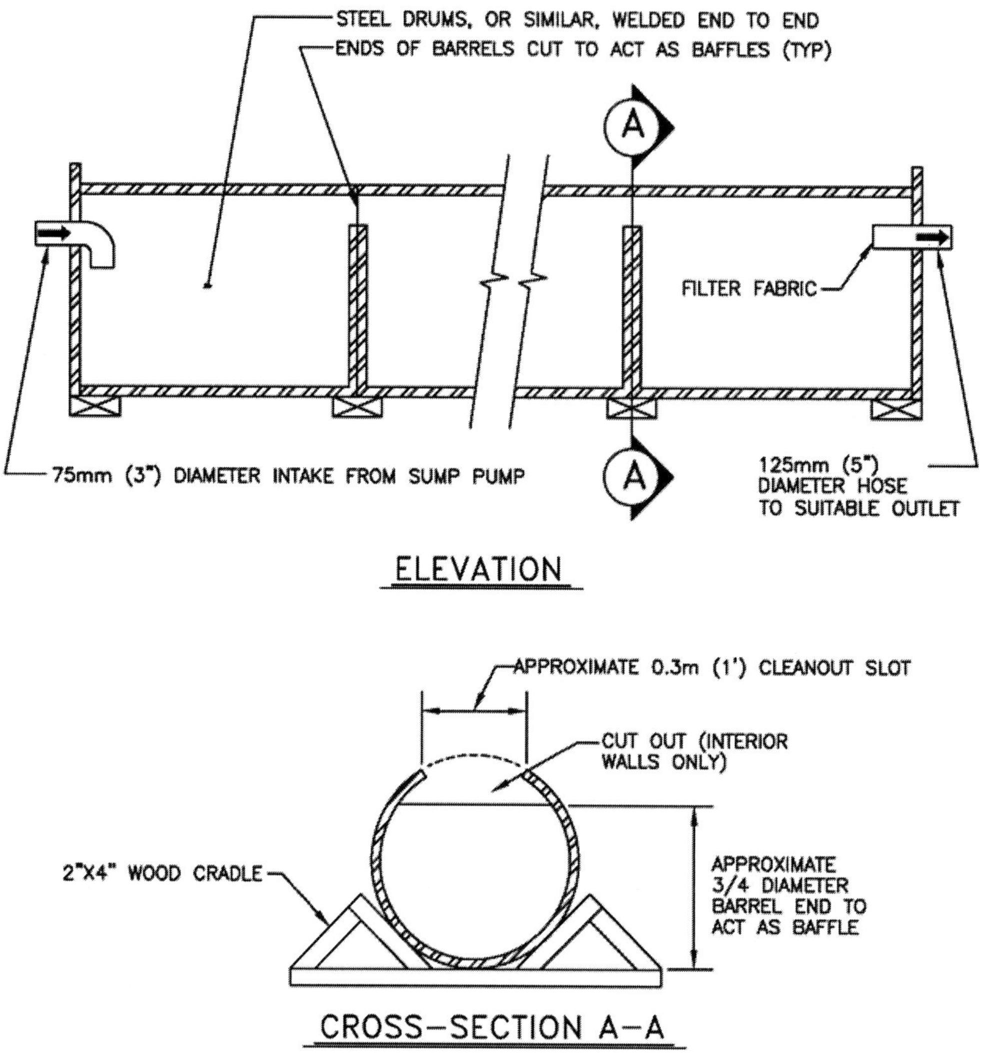

FIGURE 5-36. Sediment Tank
***Source*: Plate 3.26-1, Portable sediment tank (VDEQ 1992h), with modifications by MPCES Committee. Reproduced with permission from the Virginia Department of Environmental Quality.**

5.9 OTHER CONTROL PRACTICES

In addition to the major control practices previously described, there are a number of miscellaneous control practices that are often used on construction sites. The practices include tree protection, dust control, and chemical soil conditioners (District of Columbia 2003).

5.9.1 Tree Protection. The tree protection practice is used to provide protection of desirable trees from mechanical and other injury while the land is being developed. This practice consists of necessary protective measures, both temporary and permanent, to ensure the survival of desirable trees for shade, beautification, and vegetative cover. Fig. 5-38 provides typical details for tree protection.

5.9.2 Dust Control. Dust control practices are used for controlling blowing dust and movement on construction sites and roads. These practices are designed to prevent blowing and movement of dust from exposed soil surfaces, reduce on- and off-site damage, prevent health hazards, and improve traffic safety. Dust control practices are applicable to areas subject to dust blowing and movement where on- and off-site damage is likely without treatment.

5.9.3 Chemical Soil Conditioners. The most common chemical soil conditioner is polyacrylamide. The term *polyacrylamide*, or PAM, is a generic term that refers to a broad class of compounds. There are hundreds of specific PAM formulations, and all have unique properties that depend on polymer chain length and number and kinds of functional group substitutions along the chain. PAMs are classified according to their molecular weight and ionic charge and are available in solid, granular, liquid, or emulsion forms. PAM also is used as a flocculent aid in settling. Applying PAM to soils may improve erosion control when coordinated with other measures and practices (USDA/NRCS 2009).

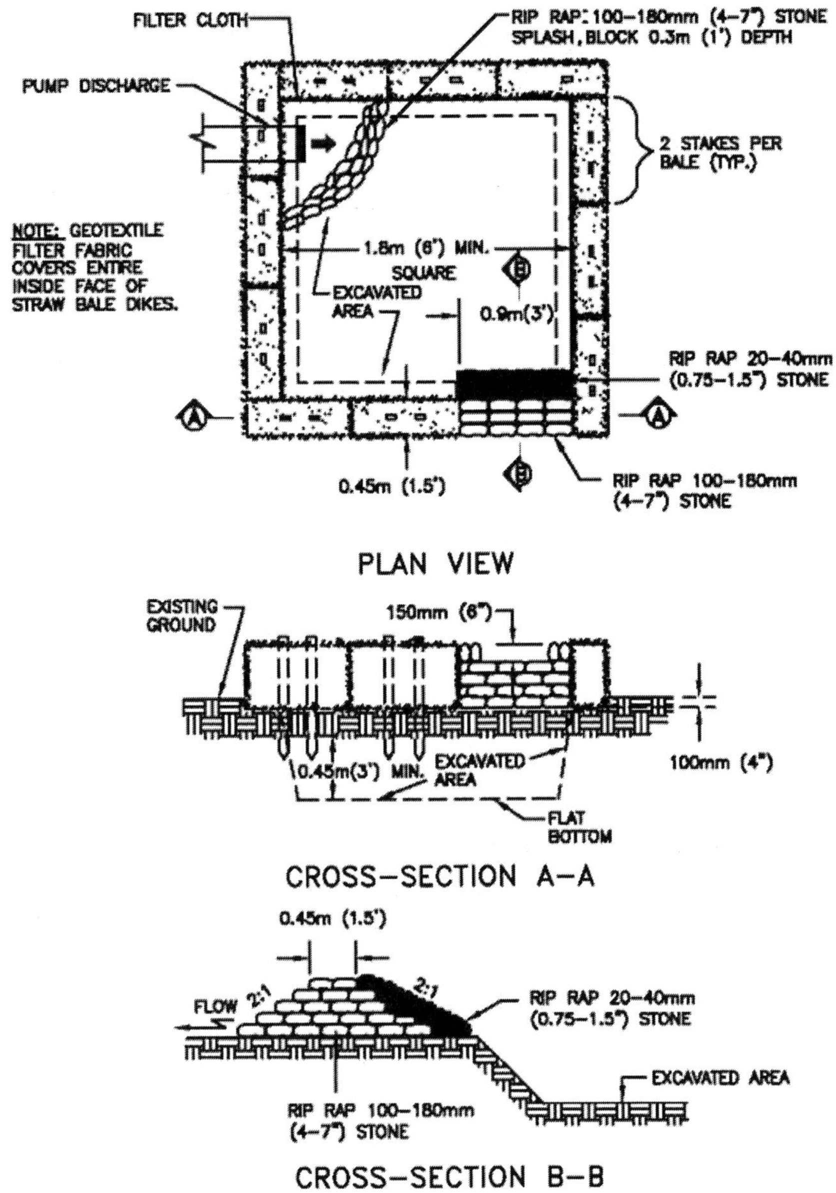

FIGURE 5-37. Dewatering Basins

***Source*: Plate 3.26-3, Straw bale/silt fence pit (VDEQ 1992i), with modifications by MPCES Committee, reproduced with permission from the Virginia Department of Environmental Quality.**

PAM's ability to prevent or reduce erosion is due to its affinity for soil particles, largely via coulombic and van der Waals attractions. These surface attractions enhance particle cohesion, stabilizing soil structure against shear-induced detachment, and transport in runoff. In a soil application, PAM aggregates soil particles, increasing pore space and infiltration capacity, resulting in reduced runoff. These larger particle aggregates are less susceptible to raindrop and scour erosion, thus reducing the potential to mobilize sediments. Use of cationic polymers requires a specialized permit (USEPA 2012).

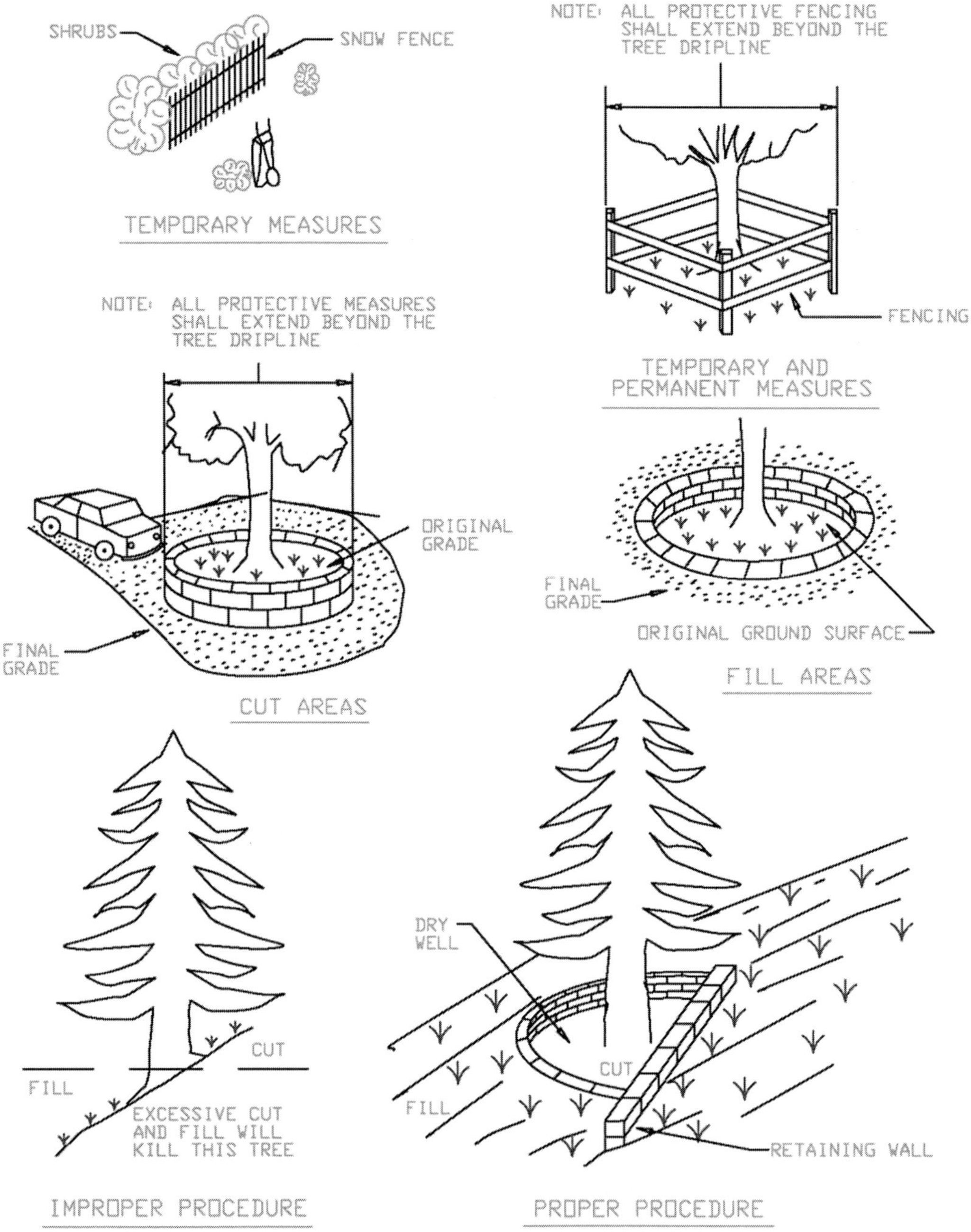

FIGURE 5-38. Tree Protection
***Source*: Fig. G233, Detail 31, Tree protection (MDE 1994r), with modifications by MPCES Committee, reproduced with permission from the Maryland Department of the Environment.**

CHAPTER 6
EROSION AND SEDIMENT CONTROL PLAN

An erosion and sediment control plan will need to be prepared and submitted to the approving agency with the project application. This plan should contain all the pertinent information for review and implementation. Most of this information was adapted from the International Erosion Control Association (IECA 2001).

6.1 COMPONENTS OF AN EROSION AND SEDIMENT CONTROL PLAN

The following components should be present and are required by the guidelines summarized in Section 3.3:

- Narrative
- Map/site plan
- Construction details, specifications, and notes
- Calculations.

6.1.1 Narrative. The narrative is a brief description of the overall strategy for erosion and sediment control. It should summarize for the plan reviewer and the project superintendent the aspects of the project that are important for erosion control and should include the following information:

- A brief description of the proposed land-disturbing activities, existing site conditions, and adjacent areas (such as creeks, stormwater conveyance, roads, and buildings) that might be affected by the land disturbance
- A description of critical areas on the site—areas that have potential for serious erosion problems, such as severe grades, highly erodible soils, and those near wetlands or water bodies
- A construction schedule that includes the date that grading will begin and the expected date of stabilization
- A brief description of the measures that will be used to minimize erosion and control sediment on the site, when those measures will be installed, and where they will be located
- A maintenance program including frequency of inspection, provisions for repair of damaged structures, and routine maintenance of erosion and sediment control practices.

6.1.2 Map/Site Plan. The map is the key item in an erosion and sediment control plan. It should show the following features:

- Existing and final elevation contours at an interval and scale sufficient for distinguishing runoff patterns before and after disturbance
- Critical areas within or near the project area, such as streams, lakes, wetlands, highly erodible soils, public streets, and residences
- Existing vegetation
- Limits of clearing and grading
- Locations and descriptive names of erosion and sediment control measures, with dimensions.

It is strongly recommended that symbols with a legend be used on the map to denote erosion and sediment control measures. Use of standardized symbols will speed up plan review and make it easier for site superintendents and inspectors to understand plans quickly. These symbols are designed to be both pictorially representative of the control measures and easy to draw. It should be noted that different erosion control maps or site plans may be required at different stages of project development, with each map or plan representing a different stage in the grading operations.

6.1.3 Construction Details, Specifications, and Notes. Construction details, often in large-scale, detailed drawings, provide key dimensions and spatial information that will not fit on the map. Other important information should also be provided. Examples are seeding and mulching specifications; equivalent opening size (EOS) and strength requirements for filter fabric; specifications for wire mesh, fence posts, and staples; installation procedures for control measures; and maintenance instructions.

6.1.4 Calculations. The calculations used to size the control measures must be included, particularly the data for the design storm (recurrence interval, duration and magnitude, drainage area, and peak intensity for the time of concentration) and the design assumptions for sediment basins and traps (design particle size, trap efficiency, discharge rate, and dewatering time). Also, calculations should be included to support the sizing of storm drain systems when an engineered design is necessary. Many regulatory agencies have developed worksheets to standardize the presentation of calculations. Use of standard worksheets typically will speed up the review process.

6.2 PREPARING AN EROSION AND SEDIMENT CONTROL PLAN

The erosion and sediment control plan is a comprehensive plan designed to address the short-term mitigation of erosion and sedimentation hazards on a disturbed site. The short-term mitigation measures are designed to address the erosion of on-site soils and the transport of sediment off site. The transport of soils off site may occur through a number of mechanisms, including being blown off site by wind, being tracked off site by vehicles, or being carried off site in stormwater runoff.

In planning the comprehensive erosion and sediment control system, the designer must answer three key questions:

(1) Where is erosion and sediment control needed?
(2) What kind of erosion and sediment control measures are appropriate?
(3) How much is enough?

To answer these questions, the designer must conduct an erosion control evaluation that

- Identifies potential erosion problems
- Develops design objectives
- Identifies and evaluates alternatives
- Selects the most appropriate erosion and sediment control measures
- Presents a comprehensive erosion and sediment control plan.

The essential steps to be followed in developing an erosion and sediment control plan are as follows:

1. Identify issues and concerns and develop goals and objectives.
2. Collect site data.
3. Analyze site data.
4. Develop site plan.
5. Identify, screen, and select best management practices.
6. Finalize the erosion and sediment control plan.

This process is described in the following sections. It is primarily designed for relatively large projects (i.e., more than one building) on several hectares (acres) or more. For very small sites, such as a single-home site, a more streamlined process may be appropriate. For example, conducting a soil particle size analysis for sizing a sediment basin for a very small site would be overkill. It would also be unnecessary to perform runoff calculations for sizing conveyances, because runoff from the site probably would be minimal, and a standard design would do the job in most cases. Many states and local jurisdictions have adopted a "standard plan," which covers sites of 1 acre or less.

6.2.1 Identify Issues and Concerns and Develop Goals and Objectives. The first step in erosion and sediment control planning is to evaluate the local, regional, and statewide issues and concerns. This evaluation will influence all other aspects of planning, including the selection of the best management practices, the time frame for implementation, and the establishment of responsibilities for various aspects of the plan. The identified issues and concerns may include the following (IECA 2001):

- The nature of the project
- Public opinion, environmental interest groups
- Contact with public agencies
- Proximity of sensitive receiving waters and other features or areas
- Regulatory environment
- Identification of high-value waters.

Once the issues and concerns are established, the goals and objectives for the project can be identified. These may include the following:

- Meeting regulations (federal, state, and local)
- Stabilizing slopes
- Reducing short-term erosion
- Reducing long-term erosion
- Reducing sediment leaving the site
- Minimizing negative public opinion
- Reducing long-term maintenance
- Maximizing the use of cost-effective solutions
- Improving aesthetics
- Enhancing the environment
- Decreasing liability exposure
- Establishing permanent vegetation
- Reducing peak stormwater runoff.

With these goals in mind, the designer can move forward with development of the plan.

6.2.2 Collect Site Data. The purpose of data collection is to gather the information on site conditions that will enable the development of an effective erosion and sediment control plan. Most of these data describe the natural environment of the site, including topography, drainage, rainfall, soils, ground cover, and adjacent areas.

It is best to collect all data in map form, if possible, and to plot data on one or more site maps at the same scale. Using a geographic information system (GIS) and/or computer aided design (CAD) software can greatly facilitate this process.

Topography—A good topographic map should form the basis of any kind of land planning, including site development planning and erosion and sediment control planning. A topographic base map can determine drainage patterns, slope lengths and angles, and locations of sensitive features on or adjacent to the site, such as water bodies, buildings, and streets. All of these are critical concerns in erosion control.

A topographic map of the site should be prepared that shows the existing contours at a suitable interval for determining drainage patterns over small areas. The contour lines must be close enough together to show which way water will flow. For most project sites, a 0.6 m (2 ft) or smaller contour interval probably will be needed.

Drainage—The drainage pattern of a site has two components: overland flow and channel flow. Both are important in erosion control. In the data collection stage, it is helpful to clearly mark all existing streams and major conveyances on the topographic base map. Major watercourses are shown as blue lines on U.S. Geological Survey topographic maps, but lesser drainage ways are also important to show. Delineating drainage ways makes determining watershed boundaries in the data analysis stage easier.

Rainfall—In erosion control rainfall data are primarily used for sizing large conveyances and sediment basins, although data can be applied to any erosion control structure through the determination of design storm and acceptable risk of failure. Rainfall frequency, intensity, and volume are the key types of data. Rainfall intensity determines the intensity value used in runoff calculations. Rainfall intensity is also a component of the R factor in the universal soil loss equation. This equation (Eq. 4.1, Chapter 4) can be used to estimate the sediment storage requirements of sediment basins. Rainfall frequency data are used for determining a "design storm."

Soils—Soils data are used to locate highly erodible areas on a site, in other words, areas where extra erosion control precautions may be needed. They also show the distribution of particle sizes in the soil, a critical factor in sizing sediment basins and traps. A high content of clay and fine silt in a soil should suggest an erosion control strategy using vegetation and mulch, as opposed to, say, straw bales and sediment basins. The latter are poorly suited for the removal of small particles, as are silt fences and conventionally sized ponds.

Soils data for many regions can be obtained from soil surveys published by the USDA/NRCS or other GIS equivalent federal mapping sources. For many projects, a soils report is specially prepared by a soils engineer. On hillside sites, many jurisdictions routinely require a soils report. If a soils report is to be prepared, it is desirable to include in it a particle size analysis for sediment basin or trap design.

Ground Cover—This term primarily refers to existing vegetation, which should be preserved to the greatest extent possible because it is the most effective form of erosion prevention. Many communities also wish to preserve trees and certain vegetation for aesthetic and other reasons. Ground cover, along with other physical characteristics of the watershed, is used to determine the C factor in the rational method, or the curve number (CN) on the USDA/NRCS methods for calculating runoff. It is also used to calculate the erosion rate in the universal soil loss equation (Eq. 4.1).

Adjacent Areas—Off-site features, such as streams, lakes, wetlands, buildings, and roads, may be particularly sensitive to erosion and sediment damage. Such areas should therefore be noted on the site map. If including the off-site features on the same map would result in an unwieldy document, one of two options can be chosen (IECA 2001): describe on the margin of the map the nature and location of the adjacent feature, or draw a smaller-scale map (vicinity map) showing the site and all the pertinent adjacent features. Many jurisdictions require designation of all receiving waters, including high-value waters, on the maps.

6.2.3 Analyze Site Data. The purpose of this step is to interpret the data collected in 6.2.2 for its significance in erosion and sediment control. This can include the following interpretations (IECA 2001):

- Identification of relevant conditions
- Identification of critical areas
- Identification of timing of soil-disturbing activities
- Evaluation of erosion potential.

These interpretations may require stating the data in a different form, such as translating a topographic map into a slope map. The result of this step is a map that highlights areas of importance in erosion and sediment control.

Drainage Areas—Understanding the site's surface water flow pattern is an essential part of this analysis. The designer must determine the following issues:

- Where will concentrated and sheet flows enter the site?
- How will runoff, both concentrated and sheet flow, travel across the site?
- Where will runoff leave the site, and will it be concentrated or sheet flow?
- How much water will flow?

Map the drainage boundaries of each of the watercourses delineated in Step 2, Collect site data, and then estimate the area of each drainage basin. If the site is larger than 2 ha (5 ac), the site may need to be subdivided into units that have smaller drainage areas. Table 6-1 provides typical drainage area limits for a number of frequently used control practices that can help the designer select the appropriate control measure for a site.

It should be noted that some of the larger ESC practices, such as the sediment basin, can control drainage areas of up to 40 ha (100 ac). However, many jurisdictions have requirements that limit the amount of area that can be disturbed at any given time, and these limits usually range from 8 ha (20 ac) to 12 ha (30 ac). In addition, the U.S. EPA, as well as a number of states, are considering reducing this limit to a smaller value of about 4 ha (10 ac).

Table 6-1. Typical Drainage Area (DA) Limits for Erosion and Sediment Control (ESC) Practices

ESC Practice	DA Limit
Pipe Outlet Sediment Trap	2 ha (5 ac)
Stone Outlet Sediment Trap	2 ha (5 ac)
Riprap Outlet Sediment Trap	4 ha (10 ac)
Sediment Basin	40 ha (100 ac)
Storm Drain Inlet Protection	0.4 ha (1 ac)
Silt Fence Culvert Protection	0.4 ha (1 ac)
Earth Dike	0.4–4 ha (1—10 ac)
Temporary Swale	2–4 ha (5–10 ac)
Dike/Swale	0.8–4 ha (2–10 ac)

Source: MDE 2011, reproduced with permission from Maryland Department of the Environment.

If grading will alter natural watershed boundaries, documenting, including mapping of the drainage boundaries that will exist after grading is completed, will need to be performed.

It should be noted that on large or multiphase projects, drainage areas and their associated outfalls may change dramatically from the preproject condition to the postproject condition and even during the life of the project. Accordingly, drainage areas may need to be delineated for several periods throughout the project.

Rainfall and Runoff— If a project requires permanent waterways and sediment basins draining large areas, rainfall frequency and intensity data will be used to calculate runoff volumes to be expected. Because these calculations must be based on specific watershed areas that drain to each planned facility, the calculations must be done at a later stage of plan development.

Slope Steepness and Slope Length—Slope steepness and slope length are critical factors in erosion control. If the slope is longer and steeper, the erosion potential is greater. If an existing long or steep slope is disturbed or a new one is created by grading, carefully designed and installed erosion control measures will be required. These measures may include benches or ditches at regular intervals, temporary vegetative stabilization, or stabilization using a covering of punched straw or other mulch material. One prime example of a steep slope is a highway embankment.

Erosion potential is closely related to slope steepness. The following slope categories can be used as guidelines for evaluating erosion potential if specific local guidance is not available:

- 0–7% slope: Low to moderate potential
- 7–15% slope: Moderate to high potential
- Over 15% slope: High to very high potential.

This slope guidance should be adjusted downward in areas with highly erodible soils or naturally steep slopes.

It is a good idea to outline on the topographic base map the aforementioned slope categories. Slopes that are over 15% and 7 to 15% slopes that are very long [over 30 m (100 ft)] should be highlighted as critical.

Soils—A soils report or soil survey covering the site should indicate soil erodibility. The K factor (see Eq. 4.1, Chapter 4) in USDA/NRCS soil surveys is an estimate of soil erodibility. Highly erodible soils should be left undisturbed. If they must be disturbed, they should be mulched and revegetated as soon as possible after grading is completed.

If the soils report gives a soil particle size distribution, check what percentage of the soil is composed of fine particles [(typically sieve #200 or 0.02 mm (0.0008 in.) or smaller)]. If a high percentage (greater than 40%) of the soil is smaller than sieve #200 or 0.02 mm (0.0008 in.), much of the suspended sediments will escape capture unless a very large sediment basin is constructed or advanced treatment techniques are employed. Remember that grading will mix topsoils with subsoils and move them around the site. If fill will be imported, this material should be analyzed also.

Ground Cover—Note any areas of critical vegetation. Vegetation on or above long or steep slopes and on highly erodible soils is particularly important for erosion control.

Adjacent Areas—Examine areas downslope from the project. Note any watercourses or other sensitive features that receive runoff from the site. Analyze the potential for sediment pollution of these watercourses and the potential for downstream channel erosion due to increased volume, velocity, and peak flow of storm runoff from the site.

6.2.4 Develop Site Plan. When a site plan is developed, erosion and sediment control should be considered along with traditional planning criteria such as economics, utility access, and traffic patterns. After analyzing the erosion hazards on site in 6.2.3, develop a site plan with erosion control in mind. Consider the following points when preparing a site plan:

Identify Sensitive Areas—In developing a site plan, one of the first steps is to identify the critical areas of the site to be developed. These would include points of ingress/egress, graded slopes, areas where vegetation is to be removed, sensitive habitat areas, and sensitive receiving water bodies. These must all be explicitly identified and addressed in the plan.

Fit Development to the Terrain—Tailor the locations of building pads and roads to the existing contours of the land as much as possible. Locate them to take advantage of the natural strengths of the site and to minimize disturbance.

Confine Construction Activities to the Least Critical Areas—Land disturbance in critically erodible areas, such as steep slopes, will require installation of costly control measures. Keeping construction out of these areas will minimize the costs.

Cluster Buildings Together—Clustering buildings minimizes land disturbance for roads and utilities and reduces erodible area. Other benefits of clustering are reduced runoff, preservation of open space, and reduced development costs.

Minimize Impervious Areas—Make paved areas, such as streets, driveways, and parking lots, as small as possible. Preserve trees, grass, and other natural vegetation. Consider paving driveways with gravel or porous paving stones. French drains, infiltration trenches, and dry wells can be used to percolate runoff from impervious surfaces into the soil if soil conditions allow. Gravel-filled trenches can be located along drip lines below roof eaves. These measures will keep runoff volumes low and minimize the need for conventional storm drains, drop inlets, and underground pipes. However, to prevent these infiltration practices from failing, it is crucial that heavy equipment be kept away from these areas to prevent soil compaction. In addition, infiltration areas should never be used for erosion and sediment control, because the sediment will clog the surface rapidly and destroy the infiltration capacity.

Retain the Natural Stormwater System—Use the natural stormwater system to convey runoff from the site wherever possible. If possible, augment the natural system with vegetated swales rather than storm sewers or concrete channels. If impervious surfaces are kept to a minimum and runoff from these surfaces is percolated into the soils on site, it may be possible, without installing channel protection measures, to use the natural stormwater system to drain a development. The cost of using the natural drainage system can be substantially lower than the cost of constructing a conventional storm drain system. Preserving the natural stormwater system also can retain a visual amenity that will enhance the value of a development.

If runoff flows will be increased by development, route these augmented flows into a stormwater conveyance system and preserve the natural stormwater system in its preexisting condition. If the stability of the natural system is upset, it may be very difficult to prevent a long-term erosion process from beginning. A constructed stormwater conveyance system can be designed to resemble a natural creek.

6.2.5 Identify, Screen, and Select BMPs. The following criteria can be used to help identify, screen, and select the most appropriate BMPs for a given site:

- Effectiveness/implementation cost
- Long-term cost (maintenance)
- Environmental compatibility
- Regulatory acceptability
- Availability
- Durability
- Longevity
- Feasibility
- Public acceptability
- Risk/liability
- Suitability for the site.

Screen the Alternatives—The purpose of screening the alternatives is to eliminate those alternatives that are not feasible for use at a particular construction site or within a watershed. Part of this screening process may be determined by the evaluation of the data collected in the first step of the process. For example, a site with predominantly clayey soils or a shallow water table could make the use of infiltration basins infeasible. The screening process may be partly influenced by the results of the quantification efforts. For example, high velocities and/or high shear stresses in unlined channels may dictate the need for channel lining.

Selection Criteria—After screening the nominated alternatives, most designers consider six characteristics in the selection process: cost-effectiveness, availability, feasibility, durability, and compatibility, as well as operation.

Cost-Effectiveness—Performance histories, both field and laboratory, should be evaluated for a number of materials and techniques and related to costs. Factors to be considered relative to costs include

- Material costs
- Preparation costs
- Installation costs
- Maintenance costs.

Factors to be considered relative to erosion control effectiveness include

- Areal density (the amount of cover on the soil surface)
- Balance between runoff/infiltration (runoff should not be increased over baseline conditions)
- Sediment reduction capabilities (ability to trap and retain sediment and prevent mobilization of soil particles)
- Reduction in water velocity (lower water velocity means less erosion potential)
- Soil texture sensitivity (some practices work better on different types of soils; other practices are unaffected by soil texture).

Availability—The material must be readily available from a local supplier or be capable of immediate shipment to the area within the time frame designated by the plans.

Feasibility—The alternative must be capable of relatively quick and easy application with minimal training required to orient crews or contractors to proper application procedures. Each material should be considered for its flexibility or applicability to a variety of field conditions. Factors to be considered relative to feasibility include

- The number of steps needed to apply the practice
- Whether hand labor or machinery is required
- Whether locally available materials can be utilized
- The time required for the practice to dry, cure, or otherwise be unaffected by rainfall.

Durability—Given the nature of the site conditions, the materials must maintain their structural integrity during installation and persist until vegetation has established effective cover.

Compatibility—Materials should be selected with regard to public acceptability and environmental sensitivity. In this regard, primary organic/biodegradable, recycled and/or photodegradable products are often preferable. Factors to be considered relative to compatibility include environmental compatibility, institutional acceptance, and visual or aesthetic impact.

An erosion control practice should be examined as to its effect on vegetation establishment, which, as previously mentioned, is considered the best form of erosion control. Characteristics that should be examined include

- Whether the practice is compatible with native plants, either introduced by seeding operations or plants material remaining on site after grading
- The practice's effect on germination and growth of plants
- Whether or not the practice requires additional fertilization of soil amendments to work
- How the practice affects soil moisture and temperature.

6.3 SITE PLAN DEVELOPMENT CONCEPTS

The possible National Transportation Product Evaluation Program's (NTPEP) standardized large-scale testing process as discussed by Kelsey (2010) could be considered before selecting the desired BMPs.

6.3.1 Evaluate Site Drainage Patterns. Determine limits of clearing and grading. Use a topographic base map that shows existing and finished contours and proposed development. The limits of the areas to be disturbed should be clearly delineated using a heavy line. This line defines the area that must be protected. The drainage boundaries defined in 6.2.3 should be reexamined to determine if they have been altered by the development plan. If so, the drainage areas that will exist after grading should be outlined on the plan. Since many control measures have a 0.4 ha (1 ac) or a 2 ha (5 ac) drainage area limit, large drainage areas should be subdivided into smaller units. Enlarged drainage areas and/or increased impervious surface within a drainage area will produce increased flows. Discharging the increased flows to existing swales and watercourses will cause channel erosion, unless or until the conveyances are adequately prepared. Determine where concentrated flows will originate on and off site, how runoff will cross the site, and where runoff will leave the site. Check for and avoid unnaturally concentrated flows in natural swales created by pipes, ditches, or berms.

The following procedures usually will ensure the development of an effective erosion and sediment control plan. The principle of developing site plans is applied in the site development process (see Section 6.2.4).

6.3.2 Minimize Exposure. The time used for grading and construction should be minimized for less exposure on large projects. Stages of construction, if possible, should move from one area to another area in order to only disturb one area in a short period of time. Apply erosion control measures as soon after land disturbance is completed as is practical and within the required time limits.

6.3.3 Retain Existing Vegetation. Retain existing vegetation wherever possible when laying out site improvements, for example trying to site buildings between existing tree clusters and build roads around trees. It is best to route construction traffic to avoid existing or newly planted vegetation. Avoid unnecessary clearing of vegetation around building pads, where construction will not be taking place. Also avoid disturbing vegetation on steep slopes or in other critical areas. Physically mark off the limits of land disturbance on the site with rope, fencing, surveyors' flags, or signs so that workers can clearly see areas to be protected. A bulldozer operator probably will not know to protect a clump of trees that is only noted on a set of plans.

6.3.4 Stabilize and Protect Denuded Areas. Reestablish vegetation on all denuded areas that will not be covered with buildings or pavement. If graded areas are to be paved or built on at a later date they must be stabilized temporarily by mulching or establishing a temporary vegetative cover on those areas. The use of asphalt-based mulches should be avoided. It is often cheaper to establish and remove a temporary cover on such an area than to repair the gullying and sediment damage that is likely to occur. Before seeding an area, make sure necessary stormwater controls are installed (see the following subsection). Plant establishment will be more successful if graded slopes are roughened or scarified before seeding. Plant establishment also must occur before so much sediment is eroded from the land surface that vegetation cannot be established.

6.3.5 Divert Runoff Away from Denuded Areas. Determine how runoff should be conveyed from the top to bottom of each drainage area. Is a diversion or a slope drain required? If so, locate it where it can intercept potentially erosive flows and route them to a well-protected outlet such as a storm drain or a lined channel. Do not allow runoff to cross a denuded or newly seeded slope or other critical areas. All conveyances and systems should be fully stabilized before allowing flows into them. If there is a significant drainage area above a cut or fill slope construct a diversion and slope drain at the top of the slope to convey the water to the bottom without causing erosion. Diversions and conveyances also can be used at the base of a disturbed slope to protect downstream areas by diverting sediment-laden runoff to a sediment trap or basin. It is often good strategy to construct a diversion or swale all the way around a disturbed area to prevent clean runoff from entering the area and also to prevent silt-laden runoff from escaping before being desilted.

6.3.6 Minimize Length and Steepness of Slopes. On long or steep disturbed or constructed slopes construct benches, diversions, or swales at regular intervals to intercept runoff. Each bench should be tilted at a gentle grade into the hill to channel the flow along the inner edge of the bench. Route the intercepted runoff to a protected outlet.

Keep runoff velocities low using the following measures:

- Minimizing flow path lengths
- Constructing channels with gentle gradients
- Lining channels with rough surfaces
- Using check dams in channels (in accordance with regulations).

6.3.7 Prepare Conveyances and Outlets to Handle Concentrated or Increased Flows. Design stormwater conveyance channels to withstand the expected flow volume and velocity from a design storm. Compute the expected discharge and velocity for both existing and newly constructed swales and for on- and off-site channels that will carry increased flow as a result of the project. Compute the shear stresses expected on the channel bottom and sides. By using these calculations, determine whether any drainage channels will require protection. If the computations indicate the runoff flow will be erosive, first determine whether a vegetative lining will be sufficient. If the expected velocity or shear stress exceeds the limit for the specified grasses, choose between rock, asphalt, or concrete linings. Remember, grass and rock linings are desirable because they keep velocities low, allow runoff to percolate into the soil, and are aesthetically pleasing. Because they resemble natural conveyances, they can enhance the appearance of a development. Also, determine whether outlet protection will be needed. Pay particular attention to transitions from pipes or paved channels to natural or unlined channels. Locate riprap aprons or other energy-dissipating devices at discharge points where erosion is likely.

6.3.8 Trap Sediment On Site. It is unnecessary, costly, and dangerous to impound runoff from large, undisturbed areas. Trap the sediment-laden runoff before it enters a stream.

Install sediment basins or traps, silt fences, or straw bale barriers below denuded areas so that runoff will be detained long enough for suspended sediment to settle out. Try to locate sediment barriers in relatively level areas or in natural depressions. A flat area at the base of a slope is a good location for a silt fence, because the runoff can slow down before reaching the barrier, and the sediment has a place to settle. Avoid placing sediment barriers where their construction would cause excessive soil disturbance. For example, excavating a sediment trap on a hillside is likely to cause more soil erosion and sedimentation than the device was intended to prevent. Also, locate sediment barriers above sensitive areas, such as streams, lakes, public streets, and adjacent properties.

Make sure there will be adequate access in wet weather for maintenance and sediment removal.

Individual lots can be surrounded by dikes to create small sediment traps called lot ponds. Gravel- or fabric-covered driveway aprons can serve as the outlets. If standing water on lots will interfere with construction activities, this type of sediment control should not be used. However, lots are often graded but not built on for months or even years. In these situations, lot ponding may be a good approach. It should be realized, however, that lot ponding may necessitate recompaction of pads prior to building construction. Consult the soils report for the soil engineer's recommendation on this issue. The size of the drainage area determines which type of sediment barrier should be used. Silt fences have a 0.4 ha/30 m (1 ac/100 ft) drainage area limit. Generally, sediment traps, depending on type, have maximum drainage areas of less than 1.2 to 4.0 ha (3 to 10 ac). A sediment basin is needed if the drainage area exceeds 4 ha (10 ac). Creating smaller subwatersheds with their own sediment basins is recommended when the drainage area is greater than 40 ha (100 ac). When a watershed is subdivided into smaller drainage areas, each with its own sediment basin, the degree of risk is likely to be substantially lower. That is, the damage that could be caused by the failure of a small basin in a small watershed is minor compared to the damage potential of the failure of a large basin in a large watershed.

Dividing a watershed into smaller drainage areas also can save money. Sediment basins are more costly than simple sediment traps to construct. In addition, sediment basins require an engineered design, whereas sediment traps typically are based on standard designs.

Sediment basins and traps commonly are located below large disturbed areas, at the lowest point in a watershed, and in swales and small conveyances. Do not locate a sediment basin in a major stream, such as one designated with a blue line on a U.S. Geological Survey topographic map, or in a wetland.

6.3.9 Inspect and Maintain Control Measures. Regardless of the BMPs selected, some follow-up is required. The instructions should specify frequency of inspections, where erosion and sediment control practices have the potential need for maintenance (particularly those used for sediment control). Although the cost of operation is seldom considered in erosion control planning, lack of maintenance is usually the primary cause of failure of an erosion and sediment control system. Factors relative to operation of installed measures include

- Maintenance and repair, particularly after storm events
- Additional fertility requirements to maintain quality of vegetation
- Irrigation, if needed, to establish and maintain effective cover.

A maintenance schedule and instructions for maintaining control measures must be developed and included in the plan. The instructions should specify where sediment dredge spoils should be placed, what spare materials (such as extra filter fabric, straw bales, stakes, and gravel) are needed, and where they should be stockpiled.

It is the responsibility of the contractor to make sure that all workers understand the major provisions of the erosion and sediment control plan. If they understand the plan, they are less likely to disturb drainage patterns and control measures, such as by running over a dike with a truck. A routine end-of-day maintenance check is strongly advised, whereas inspections are required after any storm with a prespecified amount [e.g., 6 mm (.25 in.)] of rainfall. In particular, the site superintendent should look for breaches in dikes and for erosion or sedimentation near wetlands, water bodies, discharge points, or roads. All maintenance procedures and the maintenance schedule should be specified in the plans. The plans also should remind the contractor of his or her responsibility to inform construction site workers about the erosion and sediment control features on the site.

6.4 EVALUATING AN EROSION AND SEDIMENT CONTROL PLAN

The following section is written from the perspective of a public agency plan reviewer. It describes what to look for when evaluating an erosion and sediment control plan. A checklist is helpful; however, over-reliance on checklists can give a false feeling of completeness.

6.4.1 General Considerations. *Responsibility*—It is not the responsibility of the plan reviewer to see that the plan is the best possible one. The reviewer can only ensure that the plan meets the minimum standards set by the reviewing agency and its authorizing regulations.

Communications—Encourage informal communications between the plan reviewer and the plan preparer. This will enable the reviewer to make informal suggestions that may save the developer money and the preparer time, and it may result in a more effective plan. It also will enable the preparer to explain and justify the plan. Preapplication conferences are strongly encouraged.

Required Information—Make sure all the required information has been submitted. Checklists can be used by both plan reviewers and plan preparers. However, checklists can encourage laziness. Having everything checked off does not necessarily mean that everything is in order.

Incomplete Plans—Do not review seriously incomplete plans. Send them back with a request for the missing information.

First Review—The first review should be a complete review. In subsequent reviews, deal only with items identified in the first review. It is unfair to the developer to keep injecting new requirements in subsequent reviews. However, if an important item was missed during the initial review it should be addressed immediately.

Plan Concept—The concept should be examined first, starting with the general and moving to the specific. Does the plan make sense?

Schedule—Examine the construction schedule. Does it make sense? Will the erosion and sediment controls be installed before disturbance occurs? Will disturbed and graded areas be stabilized within prescribed times?

6.4.2 Site Stormwater. Make sure the plan reviewer understands where all runoff—including roof-top runoff—comes from on and above the site, where it goes, and how it traverses the site. Has a drainage area map been provided, and is drainage area information available for each BMP structure?

Sediment Basins and Traps—Locate all sediment basins and traps, and define their contributing areas.

Erosion Control—Check the method used to prevent erosion. Hydraulic seeding and mulching may stabilize some areas adequately, but other areas, because of their proximity to sensitive features, such as watercourses, or their steepness or lack of backup protection, such as sediment basins, may need far more intensive revegetation efforts. On critical slopes, a reliable backup system for hydraulic seeding, such as punched straw, is strongly recommended. Even better, these areas should be sodded. The need for rapid vegetation establishment frequently is discovered when the planner is calculating the velocities and shear stresses of the three vegetation conditions seen during establishment: bare soil, partial establishment, and full establishment.

Channels and Outlets—Examine all conveyances where concentrated flows will occur. Be sure adequate erosion protection is provided both along channels and at channel and pipe outlets. Check the sources of runoff to be sure that all the runoff comes from undisturbed or stabilized areas or has been desilted by sediment basins or other sediment retention devices.

Miscellaneous—Look for haul roads, stockpile areas, and borrow areas. They often are overlooked and can have a substantial effect on drainage patterns. Look at all points of vehicle access to the site, and be sure mud and dirt will not be tracked onto paved streets and that sediment-laden runoff will not escape from the site at these points. Pay particular attention to watercourses and their protection.

Plan Details—Once the plan concept has been shown to be sound, check the details to be sure the concept is executed adequately in the plans.

Structural Details—Be sure that sufficiently detailed drawings of each structure (sediment basin, dike, swale, silt fence, etc.) are included so there is no doubt about locations, dimensions, or method of construction. This is particularly important if the planner is relying on standard specifications that have not been adapted to the site properly or if the planner is proposing an unconventional method of erosion control.

Calculations—See if calculations have been submitted to support the capacity and structural integrity of all structures. Were the calculations made correctly?

6.4.3 Vegetation. Look at seed, fertilizer, and mulch specifications. Check quantities and methods of application to be sure they are appropriate and consistent with local guidelines.

Be sure that general maintenance requirements and specific maintenance criteria, such as the frequency of sediment basin cleaning, are included. Are there stockpiles of spare materials (filter fabric, straw bales, stakes, gravel, etc.) to repair damaged control measures? Routine maintenance inspections should be part of the plans.

The plan must provide for unforeseen field conditions, scheduling delays, and other situations that may affect the assumed conditions.

It is worth noting that different vegetation may often only be planted during certain seasons or at certain times of the year. Given this limitation, it may be necessary to plant temporary covers only to be replaced later or to seed areas more than once. Any needed guidance concerning the timing of different plantings should be provided.

CHAPTER 7
CONTROL GUIDANCE FOR SPECIFIC TYPES OF DEVELOPMENT

Different types of land development projects may have special requirements with respect to the development of appropriate erosion and sediment control (ESC) plans. Some of these special requirements are identified and described in this section. The land development types that are addressed include

- Road construction projects
- Utility construction
- Commercial/industrial, small scale [<1 ha (3 ac)]
- Commercial/industrial, large scale [<1 ha (3 ac)]
- Residential, small scale
- Residential, large scale.

The Delaware Sediment and Stormwater Program, Division of Watershed Stewardship has published an excellent overview of this topic. The discussion that follows has been largely excerpted from DNREC (2013), with permission from the Delaware Department of Natural Resources and Environmental Control.

7.1 ROAD CONSTRUCTION PROJECTS

In developing ESC plans for road and highway construction, the plan designer shall encounter design situations ranging from new road/highway construction, existing road expansion, intersection and drainage improvements and bridge reconstruction, to shoulder widening and overlay projects. Although these types of projects differ greatly in their scope and complexity, they all share similar challenges to the ESC plan designer.

The construction project is usually linear in nature, with limited right-of-way and often crossing through multiple watersheds. Usually, only enough right-of-way is acquired to complete the required construction. This can make it difficult to maintain ESC practices once they are constructed. It may be necessary to obtain temporary construction easements for control practices, such as sediment traps. The linear nature of these projects increases the likelihood that a waterway, stream, or drainage channel could be impacted. Special attention should include both construction details and methods of construction for waterway construction.

When an existing roadway is under construction, traffic must continue around and through the work area. This could pose challenges to the ESC designer, as lanes will be shifted, interim access roads constructed, and efforts undertaken to minimize the time delay in temporary construction. The ESC plan should address the issue of highway safety from sediment leaving the construction area and include a dust control strategy. The use of tire wash facilities and street brooms may become part of the ESC plan as well. Proper ESC planning should ensure that the use of inlet protection devices does not pose a flooding hazard to existing travel lanes.

Highway construction is more dynamic than most other types of construction. Areas that are disturbed usually are not left inactive for long periods of time. This makes temporary stabilization difficult in many cases. Temporary stabilization also is challenging due to the steep slopes often used in highway embankments. Other methods of soil stabilization often must be used when organic materials, such as vegetation, cannot be used. Roadway areas that are to remain inactive and free of vegetation usually require the application of a nonorganic soil stabilization product.

Unlike private residential or commercial construction, the visual access to many of these highway projects for the public is high. Complaints could be significant if ESC is not maintained and cars travel through mud-covered roads, roadways are flooded, or dust becomes a nuisance. The public needs to be assured that reasonable measures are being taken to avoid these problems.

7.2 UTILITY CONSTRUCTION

As a gross source of sediment, utility construction is not a large contributor. This is because utility construction generally is performed in narrow rights-of-way, with most of the activity contained in trench work. But there are several unique aspects of utility work that do pose a challenge to the ESC plan designer and need to be recognized.

Large utility projects, such as sanitary sewer (SS) main installation, pose the greatest risk for ESC problems during construction. Because SS often is located in or near wetlands, flood plains and along stream corridors, the construction must be undertaken so that these resource areas are protected. Often it will be necessary for streams to be crossed by utility construction. The designer must plan for waterway construction permits, as well as prepare detailed methods for temporary stream diversions, dewatering operations, and stream crossings.

Smaller utility projects often include servicing residential development with water, sewer, telephone, electric, gas, and cable. Although the trend is toward using a common trench for several of these utilities, the installation of separate utilities can disrupt the overall sequence of construction, especially with street construction and stabilizing adjacent right-of-way areas. In a residential plan, the installation of utilities must be coordinated and ESC planned for, especially the restoration and stabilization of disturbed areas. Several rolls of erosion control matting or other materials should be on site for the immediate stabilization of small utility installations.

7.3 COMMERCIAL/INDUSTRIAL DEVELOPMENT, SMALL SCALE

When visualizing the commercial/industrial development (CID) on a small scale, it may help to think of a convenience store, gas station, fast food, retail outlet, or industrial park pad site. These small sites may range from 500 to 2,000 sq m (5,000 to 20,000 sq ft) for building footprint, associated parking and landscaped areas. The 1 ha (3 ac) limit is strictly arbitrary and was selected primarily for convenience and to reinforce the idea that this applies to small commercial site. A local jurisdiction may select another value but should keep that value relatively small.

In most instances, a bulk grading plan will not be necessary, as the site controls for ESC may be designed for the entire construction sequence. Typically, perimeter controls, such as silt fences, may be employed if the slope and slope-length values are suitable. A stabilized construction entrance at all points of ingress and egress is important and will need constant attention to maintain due to frequent traffic from trucks hauling structural building materials. Depending on the permanent stormwater design of the site, a stormwater pond may be available to use as a temporary sediment basin. If not, a temporary sediment trap may be employed, with perimeter berms directing the flow of sediment-laden water to the traps. These berms may be constructed from the topsoil stripped from the site. Used as berms around the perimeter, the topsoil does not take up room as a stockpile, which is often a problem on a small site.

Generally, the building footprint area or pad site is excavated first, with rough grading taking place around the remainder of the site. While the building area is being constructed, the storm drain system is installed, and inlet protection, if appropriate, is constructed. If the remainder of the site requires extensive grading later, temporary stabilization would be done initially. Once stabilized, this site should require simple routine maintenance until the remainder of the site area is final graded for parking and landscaping. If only minor grading of the site is required, the sequence may be such that grading and base course stone could occur early in the construction. This would also reduce the amount of bare soil exposure.

Some small commercial sites may depend on infiltration, filtration or bioretention for their permanent stormwater management. The function of these facilities often is compromised when they are utilized for sediment control, compacted by heavy equipment, or installed prematurely and allowed to become clogged with sediment. The ESC portion of the plan needs to be developed to complement the postdevelopment stormwater management strategy. Properly designed, a small-scale CID can be managed easily for proper ESC.

7.4 COMMERCIAL/INDUSTRIAL DEVELOPMENT, LARGE SCALE

Obviously, as a CID grows larger, the strategy of the ESC plan grows increasingly complex. Shopping centers, office complexes, and industrial parks all fit the heading of CID on a large scale. The designer can make this process more manageable by introducing phasing into the site planning process, even when the site is less than 8 ha (20 ac).

7.4.1 Phasing. In discussing phasing, it is important to define the clearing, grubbing, and grading stages of construction. Land-disturbing activity can be defined as a land change including but not limited to clearing, grading, excavating, transporting, and filling of land. On a wooded site, cutting down or clearing trees can be a land-disturbing activity, subject to the jurisdiction's definition. Proper sequencing involves developing a portion or phase of a site while simultaneously clearing and grading another phase. The three phases outlined below illustrate this concept.

Phase 1: Perimeter controls are established. Clearing and rough grading have been completed. It is important to note that perimeter controls should be in place prior to clearing or grubbing. The storm drain system is installed and drains to a sediment trap or to the stormwater management pond functioning as a sediment basin. All areas outside of the building area have been temporarily stabilized. The developer is now allowed to begin another phase of construction.

Phase 2: A second area of the site has been cleared of trees, and bulk grading has begun. A temporary sediment trap is in place. Diversion berms are in place directing runoff to the traps.

Phase 3: The plan review and inspection agency may allow the developer to begin clearing trees in Phase 3 without grading taking place. This may depend on whether this phase drains to a separate point. Before any stumps are removed and grading takes place, the inspection agency needs to ensure that Phase 3 is in full compliance with the plan.

Although phasing is an important tool in managing ESC activities, the plan needs to consider some flexibility among phases. If a sediment basin is to be constructed in Phase 1, for example, it may be necessary to place the excavated material at a central location, possibly in another phase. Phasing also works well if the phases are broken into separate drainage areas.

7.4.2 Two-Stage Plans. A well-designed ESC plan for large-scale CID should reflect that the site is likely mass or bulk graded. Typically, there are very few areas of these sites that will remain undisturbed, except those areas that are protected. In some cases, extreme changes in grading are necessary to ensure a relatively flat site. In other cases, multiple drainage areas should be graded to drain to one control point, or drainage areas may be divided to outlet in different directions.

During the bulk grading of a large site, the phasing is clearly the key to managing the ESC activity. However, even within a phase of construction, it may be necessary to develop two ESC plans. The first plan would be developed for the bulk grading activity. Because the stormwater or drainage collection system is not installed at this time, the ESC plan should rely on temporary berms, swales, and diversions to convey sediment-laden water to traps and basins. A second plan would be necessary when rough grading nears completion, and buildings, roads, and parking areas are under construction and now drain to the same traps and basins through an improved stormwater conveyance system. The ESC strategies are very different during the bulk grading and infrastructure development stages.

The ESC designer should be aware that sediment control facilities, even when designed and constructed properly, rarely exceed 80% removal rates for sediment. A properly designed ESC for a large-scale CID typically should involve several phases, possibly more than one ESC stage, and utilize many BMPs.

7.5 RESIDENTIAL DEVELOPMENT, SMALL SCALE

This type of residential construction involves a developer preparing an ESC plan for the roads and infrastructure. The lot grading and home construction can be handled entirely through a Standard Plan for Minor Land-Disturbing Activity. Typically, these are rural developments that involve the lots being sold individually over an extended period. This type of residential development is difficult to manage for two reasons:

1. The developer usually has fulfilled his or her responsibility once the road and infrastructure have been completed. Financial obligations are completed and guarantees are released. Typically, the stormwater management facility is completed, making it critical for the homebuilder to implement the ESC.
2. Individual lots being constructed by many different builders without the coordination of a site superintendent makes it difficult to manage drainage, runoff, soil erosion, and the resulting sediment load. If proper on-lot controls are not maintained, road swales are destroyed, culvert pipes filled with sediment, stormwater practices contaminated, and serious erosion problems could develop and be left unrepaired.

7.6 RESIDENTIAL DEVELOPMENT, LARGE SCALE

Residential development can be considered large scale when the entire site should be mass or bulk graded. There are three stages to a large-scale residential development (LSRD).

Each stage is unique with respect to erosion and sediment control, and the management of stormwater during construction. The site planning and building permit-issuing authority typically coordinates phasing of residential development. Generally, phased development is driven by the economics of home sales. It is unusual for a developer to provide infrastructure for multiple phases and provide the needed financial guarantees until income is generated through lot or home sales. Because of this, residential construction is not only a dynamic process, but multiphased projects may take years to complete. For this reason, it is important for the ESC plans to consider the necessary integration of the phases.

7.6.1 Bulk Grading Stage. As with large-scale CID, bulk or mass grading would require a separate ESC plan for that stage. As basins and traps are constructed, there is the added consideration for the planner that home lots eventually should become part of the plan, and the placement of these facilities needs to consider their long-term use. As earthwork progresses, the road areas should be "roughed" or "boxed" out if the roadway is incut or earth brought in if the road area is infill. The amount of bulk grading should depend on the earthwork balance from the site. If cuts and fills are balanced, typically the bulk grading stage should be easier to manage. Once this stage is completed, the major infrastructure stage begins.

7.6.2 Site Improvement Stage. The next stage in the LSRD involves the installation of roads, major utilities, such as sewer and water, and drainage systems. The ESC planner needs to realize that the construction of the roadway and drainage system alters the interception of stormwater runoff, and in many cases, the sheet flow occurring during the bulk grading stage is now concentrated. Energy dissipation with check dams, drop structures, and possibly turf reinforcement in swales and ditches is now necessary. Putting the base course of stone on the road as soon as possible should reduce erosion potential as well.

As the roadway cuts and fills are completed and drainage established, temporary stabilization may take place on the lot areas, and many of the roadway swales and drainage channels are ready to receive the permanent stabilization treatment. It is still early in the construction phase to activate any of the permanent infiltration/filtration facilities or systems that may have been installed. If located underground, the storm drain system must be protected to prevent soil from migrating to the infiltration system. The contributing drainage area including lot areas must be stabilized before the permanent infiltration/filtration facilities or systems are put online.

After the infrastructure is installed and before the site contractor leaves the site, the sediment basins and traps and the rest of the site should be checked to determine if maintenance is needed. Although the site should be inspected during the entire construction process, it is crucial to ensure that any major work is performed before the site contractor leaves. Often, building lots and homes are constructed by different subcontractors who may not have the heavy equipment necessary to complete the necessary maintenance.

7.6.3 Home Construction Stage. The final stage in the LSRD construction involves the home construction. The lot areas of the site that have been stabilized previously may be disturbed during the construction of the homes. Minor utilities, such as cable, electric, and telephone, generally are installed in a common trench along the road right-of-way. This installation sometimes may interfere with previously constructed ESC controls. Ideally, the utilities are installed before home construction begins and before the road right-of-way areas are stabilized. When utility installation requests are high, however, the installation priority may be tied to the number of building permits issued in a given development phase. This may necessitate the road right-of-way areas being stabilized twice.

The construction of the residential homes may be done by the land developer or a homebuilder who purchases some or all of the lots. Generally, it is easier to manage the ESC plan when the developer is the builder. Then the planner is communicating with a site superintendent who is responsible for the entire operation. Often, however, the site contractor has left the project, and it is up to the homebuilder to assume responsibility for maintaining an individual ESC plan for each home under construction. On the surface, it seems unnecessary to keep sediment on site when the development is served by a sediment basin. There are many valid reasons though for the homebuilder to maintain on-lot controls:

- The sediment basin serving the site does not function with 100% efficiency, thereby allowing a certain percentage of sediment-laden water to leave the site.
- The site developer remains responsible for cleaning out the sediment basin, maintaining roads, checking dams and any other central site controls until the contributing drainage area is stabilized.
- As homes in the development are completed, recent homeowners who are trying to establish yards and landscaping shall not be subject to sediment runoff from adjacent homes under construction. In addition, eroding soil may prevent vegetation establishment by removing the surface soil and creating a gap between sod or mats and the underlying soil.
- The homebuilder ultimately should provide a final grade and permanent lot stabilization prior to receiving a certificate of occupancy. This should be easier and less costly to accomplish if the ESC plan for the single home construction has been followed.

CHAPTER 8
CONSTRUCTION INSPECTION AND MAINTENANCE

An effective erosion and sediment control inspection program is essential for preventing and/or mitigating damage to the aquatic resources of streams and rivers by sedimentation and polluted runoff. The program ensures compliance with the federal, state, or local jurisdictions regulations by inspecting the installation of erosion and sediment control measures, monitoring and directing maintenance, and checking adherence to design standards and specifications during construction, as well as authorizing the removal of temporary controls as construction sites are permanently stabilized.

8.1 INSPECTOR GUIDANCE

This section is intended primarily for inspectors employed by regulatory agencies. Except for enforcement issues, most of the material should also apply to inspectors working for private firms. Private inspectors may also be called on to monitor and document other matters such as costs, schedule adherence, and consumption of supplies. In addition, the site contractor often has shared legal and financial responsibility for assuring that sediment pollution does not leave the site and thus should be aware of the following materials.

8.1.1 Inspector Authority and Responsibilities. Inspectors are authorized to conduct periodic inspections of all land-disturbing activity to ensure compliance with the approved plans and to determine whether the measures required in the plan are effective in controlling erosion and sedimentation resulting from the land-disturbing activity. Typically, they are responsible for and perform the following activities:

- Conduct routine inspections, as well as inspections requested by owner/contractor.
- Report on compliance, or lack thereof, with the approved soil erosion and sediment control plan.
- Report that controls are or are not installed in accordance with the standards and specifications in the approved plan.
- Report that all control measures installed are or are not being maintained until construction is completed.
- Recommend appropriate enforcement action as prescribed in the enforcement guidelines to bring sites into compliance, if needed.

The most common mistakes on job sites are poor or incorrect erosion and sediment control installation or subcontractor damage to the controls. Conveying early the importance of maintaining functionality of the installed controls to the contractor (and subcontractors) can reduce the latter.

8.1.2 Conduct. The inspector must have technical expertise in erosion prevention and sediment control. In addition, in order to achieve full compliance, the inspector must deal with people effectively. For example, disputes sometimes arise about a site violation. The person responsible for the site may disagree with the inspector's assessment of the site. Neighboring landowners may be upset and demand that corrective action be taken immediately. At times like these, property owners or their representatives may become overly emotional. The inspector must remain impartial in order to gain the respect of all parties, steer them toward resolution, and achieve compliance. No matter what the situation, inspectors need to carry out their responsibilities in a professional manner and in accordance with the rules. Inspectors must be consistent in their inspections and enforcement, handling all sites, individuals, problems, and violations in the same manner. Inspectors must maintain their integrity if they are to do the job well and be respected as professionals.

In dealing with the public, it is most important that inspectors follow proper legal procedures and remain courteous and fair. Failure to follow proper legal procedures may limit enforcement action against a violator. By showing proper courtesy and fairness, inspectors encourage mutual respect among all persons involved. Parties responsible for construction sites are less likely to violate the rules if they see the inspector as a competent professional.

The inspector, as well as all personnel associated with site erosion and sediment controls, should be experienced and trained in construction activity management, erosion and sediment controls, or other environmental training, particularly as it relates to water management or land disturbance. States and municipalities may have requirements for site inspectors, such as licensure (e.g., registered professional engineer) or certified construction training, certified erosion control, or other mandated environmental training or qualifications.

8.1.3 Compliance. The goal of an erosion and sediment control inspection program is to prevent accelerated erosion and off-site sedimentation. Although the site owner and construction contractor have the primary responsibility for managing erosion and sediment on site, the inspector is often the first person to determine if the performance standards and intent of the rule are being met. The inspector is the key person ensuring that construction sites are evaluated fairly and consistently and that the responsible party keeps the site in compliance.

The erosion and sediment control rules are performance-oriented. That is, the measures used at a construction site must be effective in controlling erosion and preventing off-site sedimentation for the site to be in compliance. Following an approved plan and installing the control measures may not be enough for a site to be in compliance with the rules. If erosion and off-site sedimentation occur, the person responsible must install additional measures to correct the problem.

The rules are also flexible, allowing the responsible parties to decide the most economical and effective means for erosion control. This encourages the use of innovative techniques and specifically designed erosion control systems. The inspector is a key individual in making this kind of performance-based rule work, because the inspector is the first person to recognize performance failures and report problems. The inspector's job includes the following evaluations:

- Determine that an erosion and sediment control plan for the site has been approved.
- Determine that all specified practices have been installed and are being maintained according to the plan.
- Determine that off-site sedimentation and turbidity are being prevented.

8.1.4 Violation Handling. If the inspector finds deficiencies, appropriate action must be taken to attain compliance. The inspector of the site plays a central role in providing details of violations and subsequent corrections. The written inspection records are the basis for enforcement and civil penalties. The inspector is the first person to determine whether the measures on the site are performing properly; thus, the inspector is the first link in enforcing the erosion and sedimentation rule. Inspectors often are called to appear at enforcement meetings or hearings as witnesses in order to document a violation.

Write a report for every inspection of a site. When writing the inspection report, remember that it is a legal document. The report must be written legibly, accurately, consistently, and in clear and concise language. The inspector is to report all violations observed each time that a site is visited, even if one has reported some of the violations on previous visits.

Write inspection reports while on the site. Writing the report on the site helps ensure items are not forgotten and allows conditions to be rechecked. Field notes are very effective in meetings. Field notes should be organized, thorough, concise, and legible. A habit of taking organized, well-written notes should aid resolution in the first enforcement case.

The inspector also should document activities thoroughly and accurately. This should be the client's best defense in the event of a violation. Good documentation should show whether the client is a "habitual offender" or a conscientious professional who has been overwhelmed by unusual events, often an important factor in determining a resolution.

8.2 SITE INSPECTION

Inspections don't "just happen." A great deal of planning and preparation goes into a proper and thorough inspection. Inspectors need to review construction plans, attend preconstruction conferences, and be knowledgeable of the law and standards. Knowing why a site is or is not in compliance is a key part of the inspector's duty.

Most noncompliance occurs because measures were not installed correctly, not maintained properly, or both. Determining the reasons why the measures are failing requires technical knowledge about the devices and how to construct them properly.

8.2.1 Inspection. An erosion control plan is designed to minimize erosion and control sedimentation. However, components of the plan may fail or the responsible party may not adhere to the plan. The responsibilities of an inspector (see Section 8.1.2) at a construction site include

- Certainty that all erosion and sediment control measures in the approved plan have been properly installed and maintained
- Evidence that erosion is being controlled
- Conviction that off-site sedimentation is being prevented
- Certainty that no turbidity in adjacent streams is being generated.

It takes time to learn how to inspect a construction site properly. Project sites are often large and can have many land-disturbing activities occurring at the same time, which can be confusing. Also, there are many considerations to keep in mind while conducting the inspection. The inspector must be familiar with the rules and many erosion and sedimentation control practices.

A proper inspection requires planning and a systematic approach. With careful preparation, the inspector can carry out the duties and work cooperatively with all responsible parties so that those involved can do their jobs efficiently.

8.2.2 Tolerances. The inspector must be reasonable regarding dimensional and performance criteria while performing inspections. This requires an understanding of the intended function of the various BMPs. Obviously, a catch basin with an opening designed to support a grate has a zero tolerance for being too small because the grate will not fit. If the opening is a 25 mm (0.5 in.) too wide, the grate will fit and still be supported by the sill or lip. If the opening is 51 mm (2 in.) too wide, the grate will fall in. This dimensional tolerance can be described as "half-inch plus, zero minus."

A stormwater pond often is designed with 305 mm (1 ft) of freeboard over the riser or spillway. High spots or slightly low spots should probably not affect the performance of the pond. In contrast, it is critical that the lip of a level spreader is installed perfectly level. In this situation, high or low spots should both have the effect of producing concentrated flows. Thus, there is almost zero tolerance, plus or minus.

Other situations are not as simple to define. The allowable (minor) amount of mud tracked or dust generated from a site may be somewhat subjective. Many factors are involved in determining performance tolerances, such as severity and frequency of infractions, efforts by the contractors, limitations of the technology and products available, and possibly several other factors. However tolerances are determined, it is essential to the integrity of the inspector and the agency that those tolerances are applied with consistency and impartiality.

8.2.3 Inspection Preparation. The first step in inspecting a project is to review plans when first submitted. This review should alert the inspector to potential problems at the site and weaknesses in the erosion and sedimentation control system design. While at the office, an inspector should look for the following items in the plan at a minimum, although there may be other items that could be included as the individual gains more experience:

- Check contour maps and available aerial photos to see how the water flows through the site. Note where water enters and leaves the site. Determine the direction of flow in the general area and in the watershed where the project is located.
- Note whether the site borders a sensitive area, such as a stream or high-quality water body. The boundary should be especially well protected from sedimentation.
- Pay particular attention to critical areas, such as step cut-and-fill slopes, stream crossings, channels, outlets of pipes and diversions, construction access routes, and highly erodible soils.

- Look for adequate access and space to maintain erosion and sediment control measures.
- Make sure that the plan provides an installation sequence for measures to control erosion and sedimentation, with measures for one phase being installed before grading of the next phase begins.
- Study the construction schedule to determine whether long periods exist between phases of construction. If so, temporary seeding or other temporary soil stabilization is required.
- Check to make sure that the plan requires all surfaces to be stabilized as soon as possible after completion of the project and within seven working days. Temporary and permanent seeding also should be specified.
- Remember that when the contractor is finished, the entire site should be stabilized so that no accelerated erosion and no off-site sedimentation occur.
- Be sure that the perimeter of the site is protected to prevent off-site sedimentation, and keep off-site runoff from flowing across highly erodible areas during construction.
- Make sure that maintenance plans are adequate and the contractor's procedure in monitoring the performance of control measures is specified. For example, it should be specified clearly whether the general contractor, subcontractor, or construction manager is to perform the inspection and/or maintenance.
- Note any proposed borrow or waste areas and proposed measures for controlling erosion and sedimentation there.
- Watch for existing areas that may not be in compliance, such as old highways and abandoned railroad rights-of-way. Those parties responsible for the land disturbance are responsible for erosion control, even if ownership of the property has changed.
- Make a list of the specific items of the plan that require closer inspection. This list can speed up the inspection and serve as a reminder to check certain important points.
- Review the erosion and sedimentation control plan; this should provide the inspector with a solid grasp of the proposed project. From the review, the inspector can identify parts of the erosion control system that may need to be strengthened and parts that should be watched carefully to see if the performance requirement is met. An individual's experience in the field and in the geographical area provides valuable assistance in the approval or revision of the submitted plan.
- Inspectors must also be familiar with the construction plans. Study these plans; identify and highlight sensitive areas, BMP placement and details, and other items of concern.

In addition, the ability to read aerial photos is important because some construction projects now use aerial photos on which to draw the construction plans. It could take some practice to be able to recognize ordinary objects from the air.

Reviewing the construction plan provides information needed for the next step of the inspection process: the preconstruction conference.

8.2.4 Preconstruction Conference. A preconstruction conference is one of the most valuable vehicles by which the inspector can address and divert many potential erosion and sedimentation problems before they become catastrophes (refer to Fig. 8-1). This conference provides an opportunity for face-to-face conversation with the responsible party and the contractor. In this way, the inspector can establish the expectations for the project and start a good working relationship with the job superintendent. While holding the conference, the inspector should keep the following suggestions in mind:

- Clarify the objectives of erosion and sediment control and inform all parties about the specific requirements for compliance in this project. Also, discuss the inspection procedures and establish communications and scheduling so that everyone knows what should be happening during the project.
- Designate a contact person for communicating concerns about erosion control. This will make future contacts much easier.
- Be sure that all parties review a copy of the approved erosion and sediment control plan so that they know what is expected and are prepared to carry out the plan.
- Inform the responsible party and the contractor that the program is performance-oriented and that the plan may need to be changed during the course of construction. Inform all parties about procedures for changing the plans.
- Try to hold the conference on the site. There, the group can walk the site and compare the plans to see if the measures are appropriate, are located properly, and can be maintained once installed. Determine areas where sediment from the sediment traps and basins can be placed and stabilized when the devices are cleaned. The site is also the best place to determine if adequate access would be available to maintain the erosion control measures.
- Discuss the schedule of clearing and grading. Emphasize that sediment control measures and perimeter control devices should be installed before the actual grading begins in order to capture sediment as it is generated. Be sure that the schedule allows for stabilizing surfaces with temporary and permanent measures during and between phases of grading and construction.
- Discuss the maintenance requirements so that the responsible party and the contractor know who is responsible for inspecting, cleaning, and repairing the measures. Regular inspection and maintenance may need to be supplemented with extra work if there is a forecast of a large storm, for example, or to clean up after a large storm.
- Establish open communications at the preconstruction conference; this provides a good foundation for the inspector's relationship with the responsible parties during the project.

8.2.5 Office Preparation. The inspector should take the time to review the plans thoroughly before going to the site, even if someone has reviewed plans previously when first submitted:

- Outline the approach for each inspection. It is necessary to know in detail the erosion control system and why each measure is specified.
- Always take a copy of the approved plans to the site for quick referral.
- Always bring the project file and necessary reporting forms.
- Always take equipment for measuring (level, tape measure, turbidity sampling kit, etc.) and documenting (camera, video recorder).
- Be sure to have all necessary personal protection, such as boots, sun, and insect protection, rain gear, water, first aid kit, radio, and whatever else is needed.

8.2.6 Site Inspection. At the construction site, the inspector should obtain answers to the following five questions:

1. Does this project have an approved permit?
2. Is the erosion and sediment control system installed as shown on the approved plans?
3. Is erosion being controlled on the site?
4. Is sediment being contained on the site?
5. Is the potential for turbidity in adjacent streams minimized?

PRECONSTRUCTION CONFERENCE CHECKLIST

Verify the following

Permits. Check that the contractor/developer has all required permits, including but not limited to federal, Federal Department of Environmental Protection, Federal Department of Transportation, water management district (local or municipal), and building permit, if required for compliance with federal, state, and local jurisdictions.

Licenses. Get the name, license number, and type of license for all contractors involved in site development.

Contacts. Get the name, phone number, and mailing address of the property owner and all contractors involved in site development.

Special Conditions. Check for special conditions attached to any permits. This could be one of the most crucial aspects of the permit.

Discuss the following

Plans and Scope of Work. Be sure that the contractor understands the plans and the tasks to be performed.

Special Conditions. Point out that any special conditions are as valid and enforceable as the permit itself, they are not optional. Make sure that everything is clearly understood.Write down any unresolved issues and follow up quickly.

Erosion and Sediment Controls. Discuss the location, proper installation, and maintenance requirements of BMPs. Examine the erosion and sediment control details.

Buffers and Natural Areas Must Be Protected. Discuss the methods used to protect these areas. Be sure that they will not be used for parking, portable toilets, material storage, waste disposal, or other unintended uses.

Tree Protection. Verify the location and type of protected trees that will remain. Discuss protection requirements and methods. Be sure that protected areas will not be used for parking, etc.

Construction Sequencing. The construction sequence will be enforced. Perimeter controls, sediment traps and basins, and necessary conveyances will be installed and stabilized before the rest of the site is cleared.

Performance Oriented Regulations. Be sure that the contractor understands that the site will be out of compliance if erosion and sediment controls fail, even if everything has been done according to plan. In that event, additional measures will be required.

Plan Changes. No construction will be allowed that differs from the plans. Major changes to the erosion and sediment control plan and any changes to the site development plan will require reapproval. Minor changes to the erosion and sediment control plan must be approved by the inspector.

Routine Inspections. Advise the contractor that the inspector will be monitoring this site for proper installation and diligent maintenance of BMPs.

Final Inspection. A final inspection will be performed when all permitted improvements have been completed. Inform the contractor about the documents which will be required at that time (e.g., operating permit, postconstruction certification, as-built drawings).

Penalties for Noncompliance. Save the best news for last.

Inspect the site

Walk or drive around the site with the contractor. Point out any potential problems on or off site. Tell the contractor what you will be looking for on your next inspections.

Affirmation

Provide the owner /contractor with a copy of your checklist, and make sure again that everything has been discussed and clearly understood. Note any clarifications, agreements, and unresolved issues. Sign and date all copies and have them do the same.

FIGURE 8-1. Sample Preconstruction Conference Checklist

If the answer to all of these questions is yes, then the site is in compliance. File an inspection report stating that the site is in compliance, and take field notes to support the inspection report. It is a good idea to keep track of the sites where the erosion and sedimentation control plans work well so that you can show others examples of compliant sites.

If the answer to any of the five questions is no, then the site is not in compliance. File an inspection report listing the items that are not in compliance. Field notes should describe precisely the noncompliance and its location. Field notes may need to be used by others, so the inspector must make them readable and understandable. The following points should help the inspector in checking for compliance:

- Bring a set of the approved plans to the site for reference. These are necessary to determine what measures make up the erosion control system and how they are to be installed and maintained.
- Take detailed, orderly field notes as the inspection takes place. This procedure should save time and possibly eliminate repeat trips to the site. Be sure that the notes are neat, concise, and complete. Remember as well that notes may be needed as evidence in court.
- Check in with the job superintendent when arriving on site so that contractors can identify the inspector and know what is being done. When possible, schedule appointments with contractors and other responsible parties in advance.
- Walk the perimeter of the site on the first inspection. This procedure should provide a fairly accurate idea of the terrain and should serve as an alert to any problems occurring from off-site water and off-site sedimentation.
- Perhaps start the inspection from the lowest point of the perimeter and work upstream through the stormwater management system. This helps raise awareness of the amount of sediment leaving the site and can help in locating its source.
- If sediment is flowing off the site, go far enough downstream to see the extent of the damage. In these situations, it is very important to document the damage. Make an estimate of the sediment volume. Photos and videotapes make very good evidence. Be sure to write the time, date, and other items in the field notes and on the inspection report.
- If turbidity is present in nearby waters, sample the stream upstream and downstream of the discharge point to provide the best possible evidence that the site is in or out of compliance.
- Bring necessary tools to measure the devices and disturbed areas in the field. Be sure that basins and traps are sized according to the plans, channels and diversions have the proper grade, and contributing areas for the control devices are no larger than those used in the design.
- Pay particular attention to the maintenance of erosion and sediment control measures. All measures require regular maintenance and may require special attention after severe storms.
- Keep in mind that when certain structural measures fail from improper installation or maintenance, more off-site sediment damage may occur than if the device had not been installed.
- Always fill out an inspection report for each trip to a site while still at the site. The pertinent inspection points are still fresh and then can be easily rechecked if any items come into question.

8.2.7 Causes of Noncompliance. When an inspector finds a site that is not in compliance, it is important for him or her to determine why. By determining the causes, solutions become more apparent. Problems of erosion and sediment control on sites fall into three categories:

1. The responsible party has not made efforts to comply with the rule.
2. There are design errors in the erosion control system or the site conditions have changed.
3. The installation or maintenance of a measure is faulty or inadequate.

8.2.8 Little or No Effort to Comply. Noncompliance in the first category is easy to spot. The responsible party may believe that the project does not come under the jurisdiction of the rule or may intentionally disregard the provisions of the rule. Quite often, these are found by inspectors while driving by the site. Therefore, inspectors should be observant in their territory.

If inspection yields a noncomplying site, the inspector must inform the responsible party that compliance is mandatory by rule. On the inspection report, note that the responsible party has been informed of the law and list the items that are not in compliance. Appropriate enforcement action then should be taken. Some of the causes of noncompliance within this category are not submitting a plan or failing to follow the approved plan.

8.2.9 Inadequate Design or Changes in Site Conditions. Violations and failures may occur because the design was inadequate or the site conditions have changed since the plan was prepared. In this event the plan needs to be revised and approved. The inspection report should note all items of noncompliance and the need for a revised plan.

Compare the original design in the plan to conditions in the field. Look for changes in the site, conditions, and construction plan. Inspectors should be able to answer the following questions when checking for violations caused by design errors and changes:

1. Are the planned measures retaining the sediment on the site?
2. Have there been modifications to the plan?
3. Are ground covers adequate for the slope and orientation of the areas to be protected, or is the slope too steep for the ground cover chosen?
4. Is the perimeter protected, given the conditions at the site?
5. Have the contributing drainage areas changed significantly, thereby potentially overloading the control measures, and are additional control measures needed?
6. Is the planned and ongoing maintenance adequate for the existing conditions?

Again, appropriate enforcement action should be taken.

8.3 INSPECTION OF EROSION AND SEDIMENT CONTROL PRACTICES

The effectiveness of an erosion and sediment control system depends on the design, installation, and maintenance of the individual practices. It is only when all three efforts have been done properly that the system functions to prevent accelerated erosion and off-site sedimentation.

Each practice has specific requirements to function properly. Inspectors must be familiar with these requirements to ensure that each practice has been designed, installed, and maintained properly. When inspecting a practice in the field, first check that the practice has been installed according to the design specifications on the approved plan. If the practice has been installed as shown on the plan, then check the appropriate section in this chapter for items that should be given special attention for each practice group.

8.3.1 Entrances and Exits. Erosion can be a special problem around all entrances and exits, access roads, and construction roads. Erosion in these places can cause mud holes, gullies, muddy pavement, dust, and complaints from neighboring landowners. Construction roads, even temporary roads, need to be stabilized to prevent erosion. An inspector should look for the following while conducting an inspection:

- Entrance and exit pads should be built with coarse gravel and stone that are sufficient to prevent tracking of sediment onto streets or other public rights-of-way and prevent the pad from sinking into the soil.
- In addition to existing controls and maintenance practices, sites displaying heavy clay soils may require the installation of a wash rack to control tracking of sediment onto roads, and this should be suggested.
- On unstable or wet soil, the stone should be spread over a layer of geotextile fabric to keep the stone from being pressed into the soil.
- Pads may need to be extended to be more effective, and this should be indicated.
- All runoff from construction roads should be diverted to sedimentation traps to retain sediments on the site.
- Pads and roads must be maintained (adding more clean stone) to ensure proper functioning.
- Public roads must be swept as required to keep them free of sediments and stone from the site.

8.3.2 Inlet Protection. Inlet protection prevents sediment from entering the storm drains and leaving the construction site. By using inlet protectors (excavated, fabric, gravel, block and gravel, or prefabricated filter bags), the designers can make use of the storm drains to discharge stormwaters during construction. The inspector should look for problem areas within each of these practices.

Excavated Drop-Inlet Protectors—If sediment has filled the excavated pool around the inlet, the contributing area for the inlet may be too large, or the inlet protection structure may not have been maintained properly. The capacity of the excavation around the drop-inlet protectors must be adequate for the contributing area. Also, the excavated area should be frequently cleaned and maintained.

Fabric Drop-Inlet Protectors—These structures frequently fail, because the posts are not set against the inlet, and the tops of the posts are not supported or braced against each other. Water should fall directly into the inlet opening, not onto the unprotected soil around the inlet box. The fabric must always be buried at the bottom to prevent undercutting and to provide structural strength. The fabric should be set a minimum of 0.6 m (18 in.) in the soil and the trench backfilled with compacted earth or crushed stone.

Drop-inlet protectors should be set low, no more than 457 mm (1.5 ft) high, to allow water to flow over them without collapsing and to prevent water from overflowing the pool behind the fabric, thus bypassing the storm inlet. In some cases, a dike may be required to prevent bypassing.

Gravel and Block, and Gravel Inlet Protectors—Gravel and block, and gravel inlet protectors should be set low, no more than 2 ft (0.6 mm) high, to prevent water from overflowing the pool and bypassing the structures. The blocks must be set against the base of the inlet for support and to prevent erosion between the blocks and the inlet. A few blocks must be set on their sides to allow the pool to drain.

The stone used for the gravel inlet protector should be large enough so that it does not wash into the inlet. The slope of the inside face of the gravel must not be too steep or the gravel will fall into the inlet.

The fine, "washed stone" must be on the outside face of the gravel inlet protector to slow the flow of the water through the larger stone. The fine stone does not filter the sediment from the water. Sediment drops out of the water because the velocity is reduced when the water is pooled behind the inlet protector.

Gravel and block, and gravel inlet protectors require flat approaches with adequate storage to allow sediment to settle. A dike may be required on the low side of the pool to prevent runoff bypassing the protector.

Prefabricated Filter Bag (Siltsack)—The grate needs to be removed and then the siltsack when sediments are within 1 ft (0.3 mm) of the grate. Proper use of heavy equipment should help avoid accident or injury. The filter bag may be replaced, or it may be emptied, cleaned, and reused.

8.3.3 Sediment Traps and Barriers. Sediment traps, basins, and barriers are used to retain sediment on the site to protect streams, lakes, drainage systems, and adjacent property. These devices are used at the outlets of channels, diversions, and other runoff conveyance measures to allow sediment-filled water to pool and for sediment to settle. These measures often are used as the last line of defense to stop sediment from leaving the site; therefore, these measures need to be inspected closely. Sediment traps should be cleaned when the sediment is one-half the design depth to maintain adequate storage volume.

Sediment Traps—Guidance for sediment trap maximum drainage area ranges from 1.2 to 4 ha (3 to 10 ac), depending on type. The size of the sediment pool must be adequate for the disturbed area. The spillway of sediment traps must be large enough to carry the design flow. The crest of the spillway should be level to allow even distribution of flow.

Geotextile fabric (filter cloth) must be installed under the outlet section to prevent undercutting. The slopes of the inside and outside faces of the outlet section must not be greater than 2:1 to prevent stone from washing away.

The earth dike forming the trap must be compacted to prevent it from failing when the pool is full. The height and top width must be adequate to hold the water in the pool. The dike must be higher than the outlet weir section or the water could wash out of the dike at its lowest point.

Sediment Basins—The size of the sediment basin must be adequate for the disturbed area. The erosion and sediment control plan should limit the drainage area to 40 ha (100 ac). Sediment basins require special attention, because their large size makes them very hazardous if they fail. Thus, it is important that sediment basins are constructed carefully to follow the dimensions, grades, elevations, pipe sizes, emergency spillway sections, and other specifications as shown on the approved plans. Large basins must be accessible to allow frequent cleaning. The sediment removed from the basins should be placed where it will not be lost off site.

Conduit must be installed and function properly. Conduit joints must be watertight and must have anti-seep collars to prevent piping along the conduit. Antiflotation weights must be used to prevent conduit movement. The soil in the embankment must be compacted to prevent piping. Hand tamping is necessary around the conduit.

Trash racks can cause failures if they are improperly designed. They should be able to catch large amounts of debris to prevent the conduit from being clogged but should not have such fine openings that they become clogged with leaves and cause water to overtop the embankment.

There should be at least 0.3 m (1 ft) of freeboard above the emergency spillway flow depth to prevent overtopping of the embankment. The emergency spillway should be large enough to carry a 25-year, 24 h storm flow safely without eroding. It should be constructed in undisturbed soil and properly stabilized.

8.3.4 Perimeter Controls (Silt Fences). Silt fences fail because they are improperly designed, installed, or maintained. Silt fences must be buried at least 203 mm (8 in.) and backfilled with compacted soil or stone to prevent undercutting. These fences must be supported adequately to prevent collapse from the pressure of the water and accumulated sediment. Silt fences should never be placed across streams, conveyances, or areas of concentrated flow. The flowing water could collapse or undermine the fence. Consider water conveyance measures for these cases. Silt fences cannot withstand flows from large areas or steep slopes. The size of the contributing area must be limited to 0.4 ha/30 m (1 ac/100 ft) of fence. In addition, silt fences require frequent maintenance. The accumulated sediment should be removed often.

8.3.5 Stream Crossings. Stream crossings must be addressed specifically and allowed by the permit. Inspect stream crossings carefully because any sediment could enter the stream directly. Debris and construction material should be removed from the stream to prevent water cutting around culverts and bridge abutments.

Culverts cause additional soil disturbance when they are installed or removed. Provisions should be made to reduce sedimentation in the stream during installation and removal of culverts. In addition, fords should be used only for shallow or intermittent streams. Use geotextile fabric covered with properly sized stone to prevent the stone from being carried downstream.

Bridges cause the least disturbance to the stream and should be used where practical.

Banks should not be filled to shorten the length of bridge required. Fills restrict the stream channel and can easily wash out.

8.3.6 Buffer Zones. The use of buffer zones to protect streams, lakes, and other bodies of water is always recommended and frequently required. Buffer zones along water bodies must be wide enough to stop all visible sediment in the first one-fourth of the buffer nearest the construction work.

Also, the use of in-stream controls, such as check dams, weirs, and the like, should be avoided.

8.4 MAINTENANCE

Maintenance of erosion control devices frequently is overlooked on many construction sites. It is one of the most critical points in preventing accelerated erosion and off-site sedimentation. The responsible party should provide for continued inspection and maintenance of erosion control practices. Maintenance for a disturbed site should be planned to continue through the life of the project.

All devices in the erosion and sediment control system should be inspected regularly, especially after storms. The erosion control plan should specify what constitutes regular inspections and proper maintenance, such as cleaning and repairs, for each practice.

Sediment traps and basins should be cleaned when the settling pools are half full. Also, contractors frequently run over diversions with heavy equipment, breaking down the dike and allowing overtopping. If a diversion must be driven by a contractor, the diversion should be stabilized with gravel and built up to the design elevation above the channel. For further maintenance, silt fences should be repaired immediately if they are damaged.

The regulations are performance-oriented. Even if practices are installed on a site according to the approved plan, the site is only in compliance when erosion and sediment are effectively controlled.

8.5 INSPECTING VEGETATION USED FOR EROSION CONTROL

More areas are protected from erosion with vegetation than with any other erosion control means. Knowing how to choose and establish the proper vegetation can prevent soil loss and sediment problems.

8.5.1 Vegetation for Erosion Control. Vegetative cover is the principal means used to stabilize soil surfaces. With the selection of the proper species and appropriate maintenance, vegetative cover provides inexpensive, long-term protection with moderate maintenance. Construction projects present a wide range of conditions for vegetation. This section describes what to look for when vegetation is used for erosion control.

A vegetative plan is one of the keys to a well-executed project. An effective plan specifies the appropriate plants for each disturbed area, describes proper soil preparation methods, and indicates when and where to plant. Vegetation should be established as soon as possible after grading. Planting should be coordinated with construction so that areas do not remain uncovered, thereby reducing unnecessary amounts of runoff and sediment.

The Site and Its Intended Uses—Vegetation works well only if the selected plant species is suitable for the climate, the soil, and the intended use for the area. Remember that at certain times of the year or under special conditions, it may be necessary to use temporary vegetation before establishing permanent vegetation. When inspecting sites that use vegetation for erosion control, the following questions should be addressed to ensure the proper use of vegetation:

- Is the plant type appropriate for the soil and the slope?
- Do the plants have fertile, well-prepared soils to grow properly (a requirement rarely met on a graded slope)? Heavy, dense subsoils may be too infertile to support certain plants. Graded slopes may be too steep or too rocky to prepare adequate seedbeds. Steep slopes may need to be sodded or covered with riprap or concrete.
- Is the plant properly chosen, given the climate and orientation of the area? It is very important that the right plant be placed in the right location for the most effective use. Is the vegetated area being maintained properly?
- Can maintenance crews reach the planted areas to provide the necessary care? Frequently the degree of maintenance required to keep a certain type of plant growing is overlooked. In addition, the responsible party must provide higher maintenance for some ornamental trees, shrubs, and grasses.
- Is the area subject to high-velocity flow?
- Do any areas, such as channels and steep slopes, require sod, riprap, or concrete linings to prevent erosion?
- Is the area going to be occasionally inundated? This would have an effect on what should be planted and where.

Seedbed Preparation—Graded areas usually are compacted and have little topsoil when planting is started. If practical, the soil should be tested so that the proper amounts of lime and fertilizer can be added. Keep in mind the essentials for plant growth: an adequate supply of nutrients, water, and air in the root zone. Slopes that have been graded often are too compacted and smooth to establish plants. It is necessary to apply fertilizer and lime (to counter acidity), prepare a proper seedbed, and roughen the surface to provide required nutrients and adequate rooting depth. Fertilizer and lime must be incorporated to a depth of 102 mm (4 in.) or more by chiseling, plowing, or rototilling. This preparation also enhances water and air infiltration to the root zone.

Proper Mulching—Seeded areas should be mulched to protect and help establish erosion control vegetation. Mulching holds the seed and fertilizer in place, protects the soil, and conserves moisture. Mulching also encourages rapid seed germination by preventing soil crusting and insulates the soil against rapid temperature changes.

Look for a proper thickness of mulch. Few areas can develop a strong growth of vegetation without mulching, and mulches are often too thin to be of much help. Mulch also needs to be well anchored to work properly. This requirement is often overlooked, causing many failures and much added expense for reseeding. On flatter slopes, mulches can be tacked by spraying on tacking agents that bind the mulch, preventing it from being washed or blown away. Crimping also works well on flatter slopes and level areas.

For steeper slopes, mulches overlaid with netting or mats can be used. Netting and mats should be anchored with long staples at the proper spacing to provide the best resistance to washing. Thicker and more durable mats should be used on steep slopes, areas that are exposed to high-velocity flows, and areas where vegetation needs more help to become established.

Maintenance of Vegetative Cover—Maintenance is the key to adequate erosion control vegetative cover. The inspector must ensure that the vegetation is protected to allow the best germination and strongest growth. Even after the vegetation has emerged, mulches and mats must be maintained to prevent plants washing away during the next rain.

Watch for areas where the mulch is too light; the mulch can blow away or wash away in the next rain. The owner/developer must have new mulch applied and must anchor it to prevent washing. Damage to seeded areas usually happens where the mulch is improperly anchored. These areas require immediate repair. The responsible party should fill the eroded area, apply new seed or plants, lime, and fertilizer, and apply an adequate layer of well-anchored mulch. If the area is in a zone where the erosion potential is high or if the practice called for in the plans is inadequate, the responsible party may need to use a heavier mat to provide more protection for the vegetation.

Look for a means of access to the vegetated areas. The responsible party cannot provide maintenance if crews cannot reach the affected areas. This is especially important for areas where high-maintenance ornamental trees, shrubs, and turf grasses have been planted.

CHAPTER 9
ENFORCEMENT

A consistent framework for enforcement of violations of soil erosion and sedimentation control regulations is an essential component of a successful erosion and sediment control program. This framework should identify and describe the enforcement remedies and procedures to be used.

9.1 GENERAL GUIDELINES

The general guidelines of enforcement herein include available remedies, selecting appropriate remedies, communications, and additional assistance in determining appropriate remedies.

9.1.1 Available Remedies. There are typically four enforcement remedies that can be used to respond to violations of federal, state, or local jurisdictions. They include, in increasing order of severity,

1. Failed Inspection and Repair Notices
2. Notices of Violation (NOV)
3. Stop Work Orders (SWO)
4. Penalties.

9.1.2 Selecting Appropriate Remedies. In determining which remedy to select, the following factors should be considered:

- The nature and severity of the problem or violation
- The urgency with which remedial activity must be taken
- Whether the violator has taken good faith measures to come into compliance
- Whether the violation is a repeat offense.

9.1.3 Communication—An Asset to Enforcement. In most cases, effective communication can foster compliance and maintain goodwill without the need to resort to the more severe enforcement remedies, such as a Notice of Infraction ("civil infractions ticket") or a SWO. It is important to develop such communication early, during the plan review and approval stage, as well as in the course of regular site inspections.

9.1.4 Additional Assistance in Determining Appropriate Remedy to Use. To ensure that the jurisdiction uniformly imposes enforcement remedies and penalties for failed inspections or violations of the erosion and sedimentation regulations, it is recommended that each jurisdiction develop written procedures for guidance to the field inspection personnel. A number of example forms developed by New Castle County, Delaware (New Castle County 2010) are provided in this section to serve as guides for jurisdictions that have not already developed such guidance documents. These include

- Erosion and sedimentation control site inspection checklist and/or inspection report (Fig. 9-1)
- NOV form (Fig. 9-2)
- SWO form (Fig. 9-3)
- Notice of Infraction (Notification of Civil Infraction) (Fig. 9-4)

9.2 FAILED INSPECTION REPORT

The site inspection report is the key inspection tool. This inspection, which typically should be conducted on a weekly basis during the site-disturbance phase of site construction, should be structured to look at all components of the site erosion and sediment control plan to verify that the controls have been installed according to the plan, that they are being maintained, and that they are functioning as intended to minimize the amount of erosion generated, as well as to ensure that all sediment is contained on the construction site.

Each component of the ESC plan is assigned a pass/fail or satisfactory/unsatisfactory score. If the report has no fail or unsatisfactory grades, the site is considered to have passed the site inspection and the owner, as well as all parties identified on the inspection form, is notified.

If a fail or unsatisfactory grade is assigned to any component of the site's ESC plan, each fail or unsatisfactory grade should be carefully considered by the site inspector. If the nature of the failure or unsatisfactory grade is a minor issue and does not render the site susceptible to off-site discharges of sediment, a repair order can be incorporated in the inspection report that allows the site owner a specified amount of time, typically 24 hours to several days, to correct the fail grade and bring the site back into compliance. The repair order should trigger a follow-up inspection by the site inspector, which may require a reinspection fee. The reinspection schedule should be set to occur shortly after the specified time to conduct the repair has elapsed.

If the nature of the failure or unsatisfactory grade is a major issue which subjects to site to potential off-site discharge of sediment, it may be necessary to move directly to a notice of violation and/or a stop work order, to bring the site into compliance with the approved ESC plan. In addition if the follow-up inspection to a repair order associated with a failed inspection report indicates that the repair order has not been addressed or resolved, the site inspector should generally proceed to either a NOV or SWO.

9.3 NOTICE OF VIOLATION (NOV)

This subsection answers the question of when to use a NOV and provides the process of issuing a NOV.

9.3.1 When to Use a NOV. The NOV is frequently the second step in a potentially escalating series of enforcement actions, although as previously indicated, if the nature of the failure is serious enough, the site inspector may resort to the NOV as the first step in the process. When used as a follow-up to a failure to implement a repair order, the NOV is an effective way to inform the site owner/contractor that:

Date:	**Time:**
SLD Application #:	**Eng. Project Phase #:**
Project/Site Name:	**Phase #:**
Location:	**Lot #s:**

Stabilized Construction Entrance — *Silt Fence*

Earth Dikes/Swales — *Inlet/Outlet Protection*

Soil Stabilization — *Sediment Traps/Basins*

Stone Check Dams — *Limit of Disturbance*

Stormwater Management Facilities — *Sequence of Construction*

Other (explain)

Site Inspected – PASS (Permits may be issued)
Site Inspected – PARTIAL PASS (See comments)
Site Inspected – FAIL (See comments)

Comments:

Fax to Builder:	*Fax #:*
Fax to Owner:	*Fax #:*
Fax to CCR:	*Fax #:*
Fax to Contractor:	*Fax #:*
Site contact: ______________________	*Phone #:*
Signature: ______________________	
Inspector: ____________________	*Phone #:*
Signature: ____________________	

FIGURE 9-1. Sample ESC Inspection Form
***Source*: Department of Land Use, New Castle County, DE, reproduced with permission.**

- Failure to address the repair order associated with a failed inspection report is not acceptable.
- A serious violation has occurred that requires immediate attention.

Information must be provided about the measures necessary to attain compliance. The period in which compliance will be achieved must be indicated. If applicable, a date for an administrative hearing of the case must be established.

The inspector should issue the NOV in a timely manner, either on the day that the violation is identified or after a second failed inspection.

9.3.2 Issuing a NOV. *Use a Standard NOV Form*—The inspector should provide the site owner/contractor with the jurisdiction's standard NOV form, including the following information (adapted, with minor revision, from New Castle County 2010).

1. Date, time, and type of inspection during which the violation was observed
2. Location of the violation, i.e., the site address
3. Name of the owner/contractor
4. Citation to the law or rule that the owner/contractor allegedly violated, as well as the action necessary to correct the violation
5. Date and time by which compliance is required
6. Notification that failure to take the measures specified will result in further enforcement action
7. Notification of the site owner/contractor's right to contest the NOV
8. Name and telephone number of the inspector.

Standard Time for Complying with a NOV—The jurisdiction may establish a standard time for complying with a NOV, such as 1 week or 15 calendar days from the date the NOV was issued. However, often the inspector may require discretion to enlarge or shorten the standard period. In determining the time for compliance, the inspector should consider the extent of the violation and the likelihood of harm to the environment from continued noncompliance (New Castle County 2010).

Service of the NOV in Person—Whenever possible, the inspector should provide the NOV directly to the site owner/contractor at the time of the inspection. At this time, the inspector also should give the operator a verbal summary of the results of the inspection, the nature of the violation(s), and the contents of the NOV. The inspector's statements and demeanor should make it clear that his or her objective is to prevent harm to health, property, and the environment, and to gain compliance with existing regulations, not take enforcement action for enforcement's sake.

Service of the NOV by Mail—In some cases, it is not possible to present the NOV to the owner/contractor in person, either because the appropriate site personnel are not available or because of a potentially adversarial or hostile atmosphere on

Date: ____________________

Location of Violation: ____________________

Permit / Project #: ____________________

Case #: ____________________

Property Owner: ____________________
____

Method of Service: Certified, First Class Mail, or Fax

Description of Violations / Code Sections Violated / Necessary Remedial Measures:

1. A stabilized construction entrance is not installed for access to the soil stockpile on lots 98 and 97. This is in violation of NCC Drainage Code 12.05.005. Directive is to install a stabilized construction entrance by (Date). Also, if work has ceased,the pile shall be stabilized within five working days of work ceasing.

DIRECTIVE:
THE RESPONDENT HAS UNTIL (DATE), TO CORRECT THE AFOREMENTIONED DEFICIENCIES PURSUANT TO §12.08.002 (A) (1) OF THE NEW CASTLE COUNTY CODE. IF YOU FAIL TO CORRECT THESE DEFICIENCIES WITHIN THE TIME PERIOD SPECIFIED, NEW CASTLE COUNTY MAY PURSUE CIVIL, CRIMINAL, AND/OR ADMINISTRATIVE REMEDIES.

If code violation(s) are not corrected within the time specified herein, the Respondent shall be required to attend a Rule to Show Cause Hearing. The fee for such a Hearing is $300.00. Additional penalties may be assessed up to $1,000.00 per day for each day a violation continues to exist.

Technical questions regarding this notice should be directed to (Site Inspector) at Tel #. ________________________.

Assistant General Manager
cc: All parties involved

FIGURE 9-2. Sample Violation Notice
***Source*: Department of Land Use, New Castle County, DE, reproduced with permission.**

the site. In those cases, the following procedures can be followed (District of Columbia 2003):

- If it is not possible to deliver the NOV in person, it can be sent to the site owner/contractor by certified mail, return receipt requested.
- If the owner/contractor fails or refuses to accept certified mail, the NOV may be served by regular first class mail as long as the address can be verified.

Reinspection to Verify Compliance—The inspector should inspect the site to verify abatement of all violations. During the visit, the inspector should summarize the reinspection on the appropriate form or inspection report, noting the status of compliance with each pending item.

9.4 STOP WORK ORDER

A stop work order (SWO) is the third level of enforcement (Fig. 9-3). It is generally considered the most stringent enforcement remedy available, as it requires cessation of all site work except that necessary to correct violations. Therefore, it may impose a significant restriction and/or cost on the site owner/contractor. For this reason, a SWO is typically only used in cases where there is an immediate environmental or health and safety threat, work is being conducted without an approved plan, or other remedies have failed to correct violations.

The SWO form typically consists of two parts: a large form to be posted at the site, and a small form for delivery to the site owner/contractor as shown in Fig. 9-3. Both forms contain the same information, as follows (New Castle County 2010):

- Name and address of the owner/contractor
- Project number or permit number for the site
- Nature, time, and place of violations
- Citation to the law or rule that the owner/contractor allegedly violated
- Corrective action required to return to compliance
- Clear notification that all work at the site is to be ceased until compliance is achieved.

DATE:________

SITE OWNER CONTRACTOR:

____

_______(Address)__________

____

RE: Stop Work Notice
Property: (Address)____
Tax Parcel(s): (Number)____
Case Number: (Number)____

Dear Respondent:

Please be advised that the attached Stop Work Notice was posted at the above-referenced property on (Date). Pursuant to this notice, all work shall stop pending the noticed Rule to Show Cause Hearing which is scheduled for (Date) at (Time) in the White Clay Creek Conference Room at the New Castle County Government Center Building located at 87 Reads Way, New Castle. If you have any questions, please contact (Site Inspector) at (Tel #) ____. Thank you.

Very truly yours,

Assistant General Manager

Attachment

cc: (All parties)

FIGURE 9-3. Sample Stop Work Letter
***Source*: Department of Land Use, New Castle County, DE; reproduced with permission.**

9.4.1 Posting SWO Forms. The inspector should post the large form in the most conspicuous location on the site so as to notify all concerned (including the public and the police department) that the site has been closed.

9.4.2 Serving of an SWO to Owner/Contractor at Site. The small form and a copy of the SWO appeals form should be presented to the site owner/contractor. If the site owner/contractor is present, the inspector should present the forms, as well as a verbal description of the nature of violations, the fact that the operation is closed, the steps necessary to correct violations, and the owner/contractor's right to appeal. The site operator should sign to confirm receipt of the SWO. The inspector should document the details by including, where possible, the names and addresses of the persons present.

9.4.3 Inability to Serve at the Site. If the owner/contractor is not present, the inspector encounters or suspects a hostile interaction, or there are other reasons why the SWO cannot be served personally, the inspector can either hand-deliver the SWO to the last known home or business address of the owner/contractor or mail the SWO to the owner/contractor's last known home or business address.

9.5 PENALTIES

The NOV and the SWO can trigger the imposition of a series of administrative fines and/or legal action. The nature of these fines or legal actions varies from jurisdiction to jurisdiction. Some representative penalty provisions are listed as follows:

- Penalty for violation of any rule, order, or condition imposed in an approved ESC plan: not less than $_______ or more than $______ for each offense. Each day that the violation continues can constitute a separate offense. Proceedings may be subject to referral or appeal to lower court.
- Penalty for intentionally, knowingly, and after written notice, violating or refusing to comply with any notice issued pursuant to existing ESC regulations: not less than $______ or more than $ _________ for each offense. Each day that the violation continues can constitute a separate offense. Proceedings may be subject to referral or appeal to the superior court.

NOTICE STOP WORK

Violation of County Code

ADDRESS OR DESCRIPTION:

TAX PARCEL NUMBER:

CODE SECTION VIOLATED:

CONDITIONS TO RESUME:

SEC 6-203 (c)(2)a.1. Hearing has been scheduled for (Responsible Party)

PLEASE NOTE: IF YOU ARE FOUND TO HAVE BEEN RESPONSIBLE FOR THE CODE VIOLATION AFTER THE RULE TO SHOW CAUSE HEARING, YOU MAY BE ASSESSED A $300.00 HEARING FEE IN ADDITION TO OTHER PENALTIES UP TO $1,000.00 PER DAY FOR EACH DAY A VIOLATION CONTINUES TO EXIST.

DATE POSTED: *PERMIT NO.* *CASE NO.*

FIGURE 9-4. Sample Notice of Infraction (Notification of Civil Infraction)
***Source*: Department of Land Use, New Castle County, DE, reproduced with permission.**

APPENDIX A
CONTROL OF EROSION AND SEDIMENT MODEL ORDINANCE

FOREWORD

This version of the Control of Erosion and Sediment Model Ordinance was developed by the Development of Regulatory Tools (DRT) Subcommittee of the ASCE/EWRI Management Practice for the Control of Erosion and Sediment (MPCES) Standards Committee (SC). New members of the DRT subcommittee provided suggestions in line with the paper prepared by Willobee and Kabbes at the International Erosion Control Association's San Diego conference in 2006. This model ordinance includes the original material (E&S Control Model Ordinance) from the EPA website that Michael Clar, chair of the MPCES SC and the original chair of the DRT Subcommittee, provided to the MPCES eRoom folder in 2005 for the development by the DRT subcommittee.

The Secretary of the DRT Subcommittee of the ASCE/EWRI MPCES SC, and past chair of the MPCES SC revised the initial draft model ordinance in 2005 to fit the format suggested by the EWRI Standards Development Council's regulatory style used for model codes, agreements, and regulations. The format used is very similar to the ASCE/EWRI Standard 40-03, *Regulated Riparian Model Water Code*, and the ASCE/EWRI Standard 33-09, *Comprehensive Transboundary Water Quality Management Agreement.* Both of these documents show the headings and articles in bold as in this **Foreword**, and the **Commentary** and **Cross References**. The commentary and cross-references are at the end of each article. This model ordinance only uses U.S. customary units.

The DRT Subcommittee members involved with the development of this standard appendix in 2009 were

Mark Willobee, *Past Chair*
Conrad G. Keyes, Jr., *Secretary*
Michael Clar, *Former Chair*
Karen Kabbes, *Member*

This appendix has been updated via the ASCE balloting process of the MPCES SC from 2009 to 2015.

PREFACE

This model ordinance borrows considerably from the U.S. Environmental Protection Agency's (USEPA) *Model Ordinances to Protect Local Resources, Model Ordinances Language, Erosion and Sediment Control* (USEPA 2006) and should be used accordingly. Additional resources were used in the development of the document.

By most accounts, the most environmentally dangerous period of development is the initial construction phase when land is cleared of vegetation and graded to create a proper surface for construction. The removal of natural vegetation and topsoil makes the exposed area particularly susceptible to erosion, causing transformation of existing drainage areas and disturbance of sensitive areas.

The USEPA model ordinance (USEPA 2006) borrowed language from the erosion and sediment control ordinance for Westchester County, New York, and also incorporates some additional features that might help prevent erosion and sedimentation and protect natural resources more fully. The model ordinance needed to be adapted to be broadly applied for several reasons. For example, some of the requirements might not be politically feasible or technically appropriate in all communities. In addition, the USEPA ordinance does not strongly correlate with the process by which building permits are granted, because the process may vary between communities.

Erosion and sediment control is widely accepted as a necessary practice, but there are certain ways to make a well-crafted ordinance more effective. First, communities need to have the staff and resources to enforce erosion and sediment control regulations; otherwise, the authority to inspect sites becomes useless. In addition, the technical manual or guidelines referenced in the ordinance needs to provide useful guidance on selecting erosion and sediment control measures; in particular, it should not include measures that are ineffective. Third, education of contractors, engineers, and designers regarding the importance and effective use of erosion and sediment controls is essential to implementing effective erosion and sediment controls.

This model ordinance contains Article 1 DECLARATION OF PURPOSE AND/OR POLICIES, Article 2 DEFINITIONS, Article 3 PERMITS, Article 4 REVIEW AND APPROVAL, Article 5 EROSION AND SEDIMENT CONTROL PLAN, Article 6 DESIGN REQUIREMENTS, Article 7 INSPECTION, Article 8 TRAINING AND CERTIFICATION PROGRAM, Article 9 ENFORCEMENT, and Article 10 SEPARABILITY.

This model ordinance is intended as an example, and for clarity, this appendix departs from the use of both SI (metric) and customary units used throughout the remainder of the standard. Consideration for use of multiple unit systems should be made when adopting this model.

ARTICLE 1 DECLARATION OF PURPOSE AND/OR POLICIES

Article 1 sets forth the purpose and/or policies that are to provide the interpretation of the *Control of Erosion and Sediment Model Ordinance* in settings for the Guidelines. This article describes in general the goals of the administering agency and how the regulators are to pursue in carrying the model ordinance into effect. While the purpose and/or goals sometimes conflict, together they express a comprehensive vision for the protection of the public within the governmental jurisdiction.

During the construction process, soil is highly vulnerable to erosion by wind and water. Eroded soil endangers water resources by reducing water quality, and causing the siltation of aquatic habitat for fish and other desirable species. Eroded soil also necessitates repair of sewers and ditches and the dredging of watercourses. In addition, clearing and grading during construction cause the loss of native vegetation necessary for terrestrial and aquatic habitat.

As a result, the purpose of this regulation is to safeguard persons, protect property, and prevent damage to the environment in ____________ (international, nation, state, county, municipality). This ordinance will also promote the public welfare by guiding, regulating, and controlling the design, construction and/or installation, use and/or operation, and maintenance of any development or other activity that disturbs or breaks the topsoil or results in the movement of earth on land in ____________ (international, nation, state, county, municipality).

Commentary: The language in this article is taken from *Model Ordinances to Protect Local Resources, Model Ordinances Language, Erosion and Sediment Control* (USEPA 2006). The jurisdictions listed have been expanded to include international through the local level of government.

Cross-references: Article 7 (Inspection)

ARTICLE 2 DEFINITIONS

The definitions in this article provide the common description of the terms for use in the articles of this document. Definitions are provided if the term appears in more than one article in the Model Ordinance, but only if the term is used in a precise sense that in some respect would not be self-evident to an ordinary speaker of American English. When the drafters of this Model Ordinance could not provide a better definition than a simple dictionary definition or when the term used is, and is expected to remain, vague rather than precise, they did not attempt to provide a definition (similar format as in *ASCE/EWRI Standard 40-03*).

2.1 *Certified Contractor*—A person who has received training and is licensed by ________________ (international, nation, state, county, municipality) to inspect and maintain *erosion and sediment control* practices.

2.2 *Clearing*—Any activity that removes the vegetative surface cover.

2.3 *Drainage Way*—Any channel that conveys surface runoff throughout the site.

2.4 *Erosion Control*—A measure that prevents erosion.

2.5 *Erosion and Sediment Control Plan*—A set of plans prepared by or under the direction of a licensed professional engineer indicating specific measures and sequencing to be used to control sediment and erosion on a development site during and after construction.

2.6 *Grading*—Excavation or fill of material, including the resulting conditions thereof.

2.7 *Perimeter Control*—A barrier that prevents sediment from leaving a site by filtering sediment-laden runoff or diverting it to a sediment trap or basin.

2.8 *Phasing*—Clearing a parcel of land in distinct phases, with the stabilization of each phase completed before the clearing of the next.

2.9 *Responsible Personnel*—Any foreman, superintendent, or project engineer who is in charge of on-site *clearing* and *grading* operations or sediment control associated with earth changes or disturbances.

2.10 *Sediment Control*—Measures that prevent eroded sediment from leaving the site.

2.11 *Site*—A parcel of land or a contiguous combination thereof, where *grading* work is performed as a single unified operation.

2.12 *Site Development Permit*—A permit issued by the municipality for the construction or alteration of ground improvements and structures for the control of erosion, runoff, and *grading*.

2.13 *Stabilization*—The use of practices that prevent exposed soil from eroding.

2.14 *Start of Construction*—The first land-disturbing activity associated with a development, including land preparation such as *clearing*, *grading,* and filling; installation of streets and walkways; excavation for basements, footings, piers, or foundations; erection of temporary forms; and installation of accessory buildings, such as garages.

2.15 *Watercourse*—Any body of water, including but not limited to lakes, ponds, rivers, streams, and bodies of water delineated by ____________________ (international, nation, state, county, municipality).

2.16 *Waterway*—A channel that directs surface runoff to a watercourse or to the public storm drain.

Commentary: Locally or regionally applicable definitions should be included in this article as necessary to clearly define words or phrases used in the articles of the ordinance. The definitions in this article are taken from *Model Ordinances to Protect Local Resources, Model Ordinances Language, Erosion and Sediment Control* (MDE 2011) and *MD Title 26.17.01.06* (COMAR 1972).

Cross-references: Article 1 (Declaration of Purpose and/or Policies), Article 3 (Permits), Article 5 (Erosion and Sediment Control Plan), and Article 6 (Design Requirements)

ARTICLE 3 PERMITS

Article 3 establishes the size and type of land-disturbing activities for which permits are required. Clearly defined permit requirements will aid in ordinance implementation and effectiveness by enabling affected parties to readily determine the applicability of the ordinance to their project(s). Public education and outreach on the permit requirements must occur at the onset of ordinance implementation and on a continual basis to aid in the development of a proactive permit program.

(A) No person shall be granted a *site development permit* for land-disturbing activity that would require the disturbance of ________________ or more square feet without the approval of an *erosion and sediment control plan* by __________________(erosion and sediment control agency).

(B) No *site development permit* is required for the following activities:

   (1) Any emergency activity that is immediately necessary for the protection of life, property, or natural resources

   (2) Existing nursery and agricultural operations conducted as a permitted main or accessory use.

(C) Each application shall bear the name and address of the owner or developer of the site and of any consulting firm retained by the applicant together with the name of the applicant's principal contact at such firm and shall be accompanied by a filing fee.

(D) Each application shall include a statement that any land clearing, construction, or development involving the movement of earth shall be in accordance with the *erosion and sediment control plan* and that a ______________ (certified contractor) shall be on site on all days when construction or *grading* activity takes place.

(E) The applicant will be required to file with ______ ______________(international, nation, state, county, municipality) a faithful performance bond, letter of credit, or other improvement security in an amount deemed sufficient by ________________________(erosion and sediment control agency) to cover all costs of improvements, landscaping, maintenance of improvements for such period as specified by ________________________ (international, nation, state, county, municipality), and engineering and inspection costs to cover the cost of failure or repair of improvements installed on the site.

Commentary: The size of the site regulated under the erosion and sediment control ordinance varies widely. The proposed Phase II of U.S. EPA's National Pollutant Discharge Elimination System (NPDES) regulates disturbances greater than 1 acre, but municipalities may regulate sites as small as 2,000 sq ft.

Municipalities may choose to exempt other activities, such as mining, from an erosion and sediment control permit, or in some cases include the exempted uses previously cited.

Some states have certified contractor programs in which contractors successfully complete a training course in basic erosion and sediment control. This person would be responsible for ensuring the regular maintenance and proper installation of *erosion and sediment control* measures.

The language in this article is taken from *Model Ordinances to Protect Local Resources, Model Ordinances Language, Erosion and Sediment Control* (USEPA 2006), but is expanded to include international governmental groups.

Cross-references: Article 2 (Definitions), Article 5 (Erosion and Sediment Control Plan), and Article 9 (Enforcement)

ARTICLE 4 REVIEW AND APPROVAL

Article 4 establishes a clearly defined review and approval process. This article will promote consistency among all agency review personnel and will aid in the applicant's project planning and scheduling.

(A) ___________________________ (*erosion and sediment control* agency) will review each application for a *site development permit* to determine its conformance with the provisions of this regulation. Within ____ days after receiving an application, ____________(erosion and sediment control agency) shall, in writing:
  (1) Approve the permit application
  (2) Approve the permit application subject to such reasonable conditions as may be necessary to secure substantially the objectives of this regulation, and issue the permit subject to these conditions, or
  (3) Disapprove the permit application, indicating the reasons and procedure for submitting a revised application and/or submission.

(B) Failure of the _________________________(*erosion and sediment control* agency) to act on an original or revised application within _____days of receipt shall authorize the applicant to proceed in accordance with the plans as filed unless such time is extended by agreement between the applicant and ____________________ (*erosion and sediment control* agency). Pending preparation and approval of a revised plan, development activities shall be allowed to proceed in accordance with conditions established by ________________ (*erosion and sediment control* agency).

Commentary: The time and provisions provided in this article should be set based on detailed consideration of available staff resources and the expected workload. The time and provisions should not be set in a manner as to prevent the agency to meet the requirements of its own ordinance. The language in this article is taken from *Model Ordinances to Protect Local Resources, Model Ordinances Language, Erosion and Sediment Control* (USEPA 2006) and has been expanded to include international governmental groups.

Cross-references: Article 2 (Definitions), Article 3 (Permits), Article 6 (Design Requirements), and Article 9 (Enforcement)

ARTICLE 5 EROSION AND SEDIMENT CONTROL PLAN

Article 5 provides the general framework for the development of the *erosion and sediment control plan.* This article also establishes the *erosion and sediment control plan* modification/amendment process.

(A) The *erosion and sediment control plan* shall include the following:
  (1) A natural resources map should identify soils, forest cover, and resources protected under other chapters of this code.
  (2) A sequence of construction of the development site should include stripping and clearing; rough grading; and construction of utilities, infrastructure, and buildings. It also should include final grading and landscaping. Sequencing shall identify the expected date on which clearing will begin, the estimated duration of exposure of cleared areas, areas of clearing, installation of temporary *erosion and sediment control* measures, and establishment of permanent vegetation.
  (3) All *erosion and sediment control* measures necessary to meet the objectives of this local regulation throughout all phases of construction and after completion of development of the site should be present. Depending on the complexity of the project, the drafting of intermediate plans may be required at the close of each season.
  (4) Seeding mixtures and rates, types of sod, method of seedbed preparation, expected seeding dates, type and rate of lime and fertilizer application, and kind and quantity of mulching for both temporary and permanent vegetative control measures should be provided.
  (5) Provisions for maintenance of control facilities should include easements and estimates of the cost of maintenance.

(B) Modifications to the plan shall be processed and approved or disapproved in the same manner as Article 4 of this regulation, may be authorized by_______________ (*erosion and sediment control* agency) by written authorization to the permittee, and shall include
  (1) Major amendments of the *erosion and sediment control plan* submitted to________________ (*erosion and sediment control* agency)
  (2) Field modifications of a minor nature.

Commentary: The map should be at a scale no smaller than 1 in. = 100 ft.

Effective erosion and sediment controls must include both erosion control measures and sediment control measures. Planning for erosion and sediment control should begin at the onset of the project planning process and continue throughout the life of the project. The language in this article is taken from *Model Ordinances to Protect Local Resources, Model Ordinances Language, Erosion and Sediment Control* (USEPA 2006) with minor revisions.

Cross-references: Article 2 (Definitions), Article 3 (Permits), Article 4 (Review and Approval), and Article 6 (Design Requirements)

ARTICLE 6 DESIGN REQUIREMENTS

Article 6 describes the design requirements for the development of *erosion and sediment control plans*. This article also describes the standards that must be achieved through the proper design and implementation of the *erosion and sediment control plan*.

(A) *Grading*, *erosion control* practices, *sediment control* practices, and waterway crossings shall meet the design criteria set forth in the most recent version of________________ (erosion and sediment control manual) and shall be adequate to prevent transportation of sediment from the *site* to the satisfaction of______________*(erosion and sediment control* agency). Cut and fill slopes shall be no greater than 2:1, except as approved by ________________ *(erosion and sediment control* agency) to meet other community or environmental objectives.

(B) *Clearing* and *grading* of natural resources, such as forests and wetlands, shall not be permitted, except when in compliance with all other chapters of this code. *Clearing* techniques that retain natural vegetation and drainage patterns, as described in __________________ *(erosion and sediment control* manual), shall be used to the satisfaction of ___________________ *(erosion and sediment control* agency).

(C) *Clearing*, except that necessary to establish *sediment control* devices, shall not begin until all *sediment control* devices have been installed and have been stabilized.

(D) *Phasing* shall be required on all *sites* disturbing greater than _____ acres, with the size of each phase to be established at plan review and as approved by_____________________ *(erosion and sediment control* agency).

(E) *Erosion control* requirements shall include the following:
   (1) Soil stabilization shall be completed within 5 days of *clearing* or inactivity in construction.
   (2) If seeding or another vegetative *erosion control* method is used, it shall become established within 2 weeks or the _____________________ *(erosion and sediment control* agency) may require the *site* to be reseeded or a no vegetative option employed.

(F) Special techniques that meet the design criteria outlined in_____________________ *(erosion and sediment control* manual) on steep slopes or in drainage ways shall be used to ensure stabilization.
   (1) Soil stockpiles must be stabilized or covered at the end of each workday.
   (2) Entire *site* must be stabilized at the close of the construction season, using a heavy mulch layer or another method that does not require germination to control erosion.
   (3) Techniques shall be employed to prevent the blowing of dust or sediment from the *site*.
   (4) Techniques that divert upland runoff past disturbed slopes shall be employed.

(G) *Sediment control* requirements shall include
   (1) Settling basins, sediment traps, or tanks and *perimeter controls*
   (2) Settling basins that are designed in a manner that allows adaptation to provide long-term stormwater management, if required by _______________ *(erosion and sediment control* agency)
   (3) Protection for adjacent properties by the use of a vegetated buffer strip in combination with *perimeter controls*.

(H) Waterway and watercourse protection requirements shall include
   (1) A temporary stream crossing installed and approved by______________________________ (approving agency, e.g., Waterways Division, ESC agency) if a wet *watercourse* will be crossed regularly during construction
   (2) *Stabilization* of the *watercourse* channel before, during, and after any in-channel work
   (3) All on-site stormwater conveyance channels designed according to the criteria outlined in _____ ___________________ *(erosion and sediment control* manual)
   (4) Stabilization adequate to prevent erosion located at the outlets of all pipes and paved channels.

(I) Construction site access requirements shall include
   (1) A temporary access road provided at all *sites*
   (2) Other measures required by__________________ (erosion and sediment control agency) to ensure that sediment is not tracked onto public streets by construction vehicles or washed into storm drains.

(J) *Erosion control* practices included in the *erosion and sediment control plan* designed in a manner as to limit the soil loss from the disturbed *site* to ______tons/acre/year.

(K) The *erosion and sediment control plan* designed and implemented in a manner as to not allow the concentration of total suspended solids (TSS) in construction site discharges to exceed ________mg/L at any given time during the project life.

Commentary: Although many communities encourage *phasing*, few actually require it.

Phasing construction can reduce erosion significantly when well designed (see Claytor 1997). Numerical standards regarding the time to stabilization will vary. In particular, the time to establish seeding will depend on the climate. Dust control is most important in arid regions of any region.

The soil loss standard can be included to place a greater emphasis on the erosion control portion of the *erosion and sediment control plan* rather than on the heavy dependence on *sediment control* portion, which generally is seen in plans and on *sites*. Dane County, Wisconsin, for example, limits the soil loss rate from construction sites to 7.5 tons/acre/year. As part of the technical manual referenced in its ordinance, the Dane County Land Conservation Department has developed a locally applicable spreadsheet based on the Universal Soil Loss Equation (USLE) that allows users to evaluate erosion control practices and combinations of practices for meeting the soil loss standard (Dane County 2007). The *erosion and sediment control* agency would have to develop a similar locally applicable tool to facilitate the use of such a standard. Contact the local U.S. Department of Agriculture, Natural Resources Conservation Service office for possible assistance.

The TSS limit should be established per state and/or local water quality regulations. A majority of the language in this article is taken from *Model Ordinances to Protect Local Resources, Model Ordinances Language, Erosion and Sediment Control* (USEPA 2006).

Cross-references: Article 2 (Definitions) and Article 5 (Erosion and Sediment Control Plan)

ARTICLE 7 INSPECTION

Article 7 describes the inspection requirements for both the erosion and sediment control agency and the permittee or his or her agent and establishes coordination and communication between the two parties at the onset of the construction phase of the project. This article also establishes the *erosion and sediment control* agency's right to access the site for inspection purposes.

(A) The permittee or his or her agent and *site* contractors shall schedule a preconstruction meeting with ______ ________________ (*erosion and sediment control* agency) or its designated agent at least 1 week prior to the start of construction.

(B) ____________________ (*erosion and sediment control* agency) or designated agent shall make inspections as hereinafter required and either shall approve that portion of the work completed or shall notify the permittee wherein the work fails to comply with the *erosion and sediment control plan* as approved. Plans for *grading*, stripping, excavating, and filling work bearing the stamp of approval of the ____________________ (*erosion and sediment control* agency) shall be maintained at the *site* during the progress of the work. To obtain inspections, the permittee shall notify_________________ (*erosion and sediment control* agency) at least 2 working days before

   (1) Start of construction
   (2) Installation of sediment and erosion measures
   (3) Completion of *site clearing*
   (4) Completion of rough *grading*
   (5) Completion of final *grading*
   (6) Close of the construction season
   (7) Completion of final landscaping.

(C) The permittee or his or her agent shall make regular inspections of all control measures in accordance with the inspection schedule outlined on the approved *erosion and sediment control plan*. The purpose of such inspections will be to determine the overall effectiveness of the control plan and the need for additional control measures. All inspections shall be documented in written form and submitted to_________________ (*erosion and sediment control* agency) at the time interval specified in the approved permit.

(D) ______________________ (*erosion and sediment control* agency) or its designated agent shall enter the property of the applicant as deemed necessary to make regular inspections to ensure the validity of the reports filed under Section C and to monitor and sample site discharges as deemed appropriate by ______________________ (*erosion and sediment control* agency).

Commentary: The Certified Inspector Program in Delaware allows developers to hire an inspector who has passed a state licensing program. This person inspects the site at regular intervals and files reports to the *erosion and sediment control agency*. The agency would then be responsible for spot checks on these reports.

A majority of the language in this article is taken from *Model Ordinances to Protect Local Resources, Model Ordinances Language, Erosion and Sediment Control* (USEPA 2006)

Cross-references: Article 2 (Definitions), Article 3 (Review and Approval), and Article 6 (Design Requirements)

ARTICLE 8 TRAINING AND CERTIFICATION PROGRAM

(A) The __________________ (*erosion and sediment control* agency) shall require certification of *responsible personnel* as established by this article. *Responsible personnel* shall obtain certification by completing the__________ ________ (*erosion and sediment control* agency) approved training program. Enrollment of existing and future *responsible personnel* is the responsibility of employers. Response to a _________________(erosion and control agency) notice of training and certification (in accordance with the provisions of section B) shall serve as an application for training. The ________________ (*erosion and sediment control* agency) shall notify employers of *responsible personnel* as to the date and location of training programs for attendance by *responsible personnel* and other interested persons.

(B) After ______________________, employers of *responsible personnel* may receive interim certification for *responsible personnel* during the period before completion of the ______________________ (*erosion and sediment control* agency) training program by submitting the enrollment form to the _________________ (*erosion and sediment control* agency). Interim certification shall be valid until the scheduled date of attendance for training of *responsible personnel*. These enrollment forms are available from the ______________________ (*erosion and sediment control* agency) and the soil conservation districts. Employers shall submit the enrollment form to the _________________ (*erosion and sediment control* agency) within 30 days of receipt of the notice of training. At a minimum, the following items of information are to be submitted:

   (1) Employer's name
   (2) Employer's address
   (3) Employer's telephone number
   (4) Name of *responsible personnel*
   (5) The most advantageous location (international, country, state, county, or municipality) for each *responsible personnel* to attend a training program.

(C) The ______________________ (*erosion and sediment control* agency) may approve training programs other than those developed by the _____________ (*erosion and sediment control* agency). Unless otherwise approved by the ______________________ (*erosion and sediment control* agency), an acceptable training program shall include the following:

   (1) Educational materials covering the following topics:

(a) Ecological and resource values of the waters of the location
(b) Benefits of proper and effective *erosion and sediment control* implementation and maintenance
(c) Purpose and provisions of *erosion and sediment control* laws, ordinances, and regulations
(d) Description of sediment as a pollutant
(e) Process of
    (i) Erosion
    (ii) Sediment transport
    (iii) Sediment deposition.
(f) Proper implementation of *erosion and sediment control*
(g) Recognition and correction of improperly implemented *erosion and sediment controls*
(h) Proper maintenance of *erosion and sediment controls*
(i) Responsibilities of supervisory and enforcement personnel.

(2) Criteria for qualifying as a training instructor shall include knowledge of the principles of
(a) Erosion
(b) Sediment transport
(c) Sediment deposition
(d) *Erosion and sediment control* technology
(e) *Erosion and sediment control* implementation
(f) *Erosion and sediment control* maintenance
(g) Local government *erosion and sediment control* ordinances.

(D) Instructors of training programs shall submit the following information to the ___________________ (erosion and sediment control agency):
(1) Names, addresses, and telephone numbers of *responsible personnel* who complete a training program
(2) Card certification numbers issued to *responsible personnel*
(3) Employer's name and address.

(E) *Responsible personnel* who complete the ___________ (erosion and sediment control agency) training program shall receive the _________________ (erosion and sediment control agency) Certification of Training from the instructor. This Certificate of Training is a card with a certification number to be carried by the *responsible personnel.*

(F) *Responsible personnel* certified by the ______________ (erosion and sediment control agency) shall retain their certification status when changing employment.

(G) A Certificate of Training is valid for a 3-year period and is automatically renewed unless the ______________ __________ (erosion and sediment control agency) notifies the certification holder that additional training is required.

Commentary: Through training and experience, certified professionals have the ability to work with developers, contractors, and regulators to determine if the *erosion and sediment control plan* is being properly implemented or if the plan needs to be altered based on the given site conditions. An example of an increasingly recognized certification program is the Certified Professional in Erosion and Sediment Control (CPESC) designation provided through CPESC, Inc. (www.cpesc.net). This article is taken from *MD Title 26.17.01.06* (COMAR 1972). It is suggested that appropriate state training courses be prescribed. For example, the *Contractor's Training Course and Certified Construction Reviewer's Course* from Delaware were recommended by Evans (2004). State training programs for inspectors also have been undertaken in other states.

Cross-references: Article 2 (Definitions) and Article 7 (Inspections)

ARTICLE 9 ENFORCEMENT

Article 9 establishes the appropriate course of action to be taken and the penalties to be levied by the erosion and sediment control agency in the event of finding a person, or persons, in violation of the terms of the site development permit or for failure to obtain a site development permit for land-disturbing activities covered under the terms of the ordinance.

(A) Stop-Work Order; Revocation of Permit
In the event that any person holding a *site development permit* pursuant to this ordinance violates the terms of the permit or implements site development in such a manner as to materially adversely affect the health, welfare, or safety of persons residing or working in the neighborhood or development site so as to be materially detrimental to the public welfare or injurious to property or improvements in the neighborhood, _______________ (erosion and sediment control agency) may suspend or revoke the *site development permit.*

(B) Violation and Penalties
No person shall construct, enlarge, alter, repair, or maintain any *grading,* excavation, or fill, or cause the same to be done, contrary to or in violation of any terms of this ordinance. Any person violating any of the provisions of this ordinance shall be deemed guilty of a misdemeanor, and each day during that any violation of any of the provisions of this ordinance is committed, continued, or permitted shall constitute a separate offense. Upon conviction of any such violation, such person, partnership, or corporation shall be punished by a fine of not more than $______ for each offense. In addition to any other penalty authorized by this article, any person, partnership, or corporation convicted of violating any of the provisions of this ordinance shall be required to bear the expense of such restoration.

Commentary: Specific penalties will vary between communities and should reflect enforceable penalties given the political realities of a jurisdiction. A majority of the language in this article is taken from *Model Ordinances to Protect Local Resources, Model Ordinances Language, Erosion and Sediment Control* (USEPA 2006).

Cross-references: Article 2 (Definitions)

ARTICLE 10 SEPARABILITY

The provisions and articles of this ordinance shall be deemed to be separable, and the invalidity of any portion of this ordinance shall not affect the validity of the remainder.

Commentary: This article allows any of all of the provisions of this ordinance to be used in the rules and/or regulations of the governmental body and the validity of the used provisions would need to be cross-referenced accordingly.

Cross-references: Articles 1 through 9, if they are all used; otherwise the appropriate ones should be listed at this location.

APPENDIX B
U.S. REGULATIONS AND PROGRAMS

There has been no single piece of federal legislation drafted specifically to address problems associated with erosion and sedimentation. However, since passage of the Federal Clean Water Act in 1972, numerous federal programs that affect the management and protection of water resources have been implemented by a wide variety of federal agencies. Most of these programs address erosion and sedimentation problems through the management of stormwater, especially with respect to reducing the impacts of stormwater on aquatic systems.

This appendix provides information from two primary sources:

- *Delaware Erosion and Sediment Control Handbook.* (2013). Delaware Sediment and Stormwater Program, Division of Soil and Water Conservation, Department of Natural Resources and Environmental Control, Dover, DE.
- International Erosion Control Association (IECA). (2001). Workbook: How to Select, Install, and Inspect Construction Site Erosion and Sediment Control BMPs for NPDES Storm Water Permit Compliance. IECA, Denver, CO.

Following is an overview of several federal programs in effect at the time this appendix was prepared, which either directly or indirectly impact the management of stormwater and other nonpoint sources of pollution (USEPA 1993).

B.1 U.S. ENVIRONMENTAL PROTECTION AGENCY (EPA)

B.1.1 Clean Water Act of 1972. The Clean Water Act of 1972, Section 104, Water Quality Cooperative Agreements, allows establishment of national programs for the prevention, reduction, and elimination of pollution. Some program activities include special water quality studies, investigations of pollution control techniques, river corridor watershed management planning, and pilot and demonstration projects to implement NPDES related activities, especially those related to the control of stormwater and Combined Sewer Overflow (CSO) discharges. Special programs that are funded through Section 104 include:

- **Near Coastal Waters**, which is a program to improve the environmental conditions of near coastal waters through the use of a watershed management approaches. Eligible activities include the development and implementation of regional strategies in targeted areas.
- **State Wetlands Program**, which is designed to increase the ability of state programs to protect wetland resources. Eligible activities include the development of new state wetland protection programs or the refinement of existing programs, watershed protection demonstration projects, state wetland conservation plans, and Section 404 program assumption assistance.
- **Section 104(g)** is intended to encourage the establishment or enhancement of state small community outreach programs.
- **Section 106, Water Pollution Control** is for the administration of programs for the prevention, reduction, and elimination of water pollution. This is the primary federal grant funding source for state water quality management programs. Eligible activities include nearly all aspects of the prevention and abatement of surface and ground water pollution (planning, monitoring, permitting, enforcement, training, public education, and technical assistance).
- **Section 205(j)(1)** authorizes water quality management planning programs by states. Eligible activities include identifying pollution control methods to protect and restore water bodies and monitoring programs conducted statewide or in targeted watersheds. This is being replaced by Section 604(b).
- **Section 303, Water Quality Standards and Implementation Plans** requires states to develop, adopt, and enforce water quality standards, which are at least as stringent as those adopted by the EPA, that will protect, maintain, and restore the chemical, physical, and biological integrity of all waters. This requires regular review and, if needed, revision of water quality standards; the establishment, allocation, and enforcement of total maximum daily loads for certain water bodies; and a continuous state water planning process.
- **Section 304, Information and Guidelines** requires EPA and the states to establish water quality criteria and effluent guidelines for a wide variety of substances, especially hazardous and toxic ones. This requires states to develop a list of waters needing control strategies for toxic and other pollutants.
- **Section 305, Water Quality Inventory** requires EPA to biennially report to Congress on the environmental quality of the nation's water resources including an identification of which waters are or are not meeting their designated uses. States develop State Water Quality Assessment reports [305 (b) reports] and submit them to EPA, which uses them as the basis for its report.
- **Section 314, Clean Lakes Program** is for the establishment of projects and programs to control pollution sources to lakes and protect/restore the quality of lakes. Eligible activities include identification and classification surveys of all publicly owned lakes, state lake water quality monitoring and assessment, and public education. Lake restoration projects typically include three phases: Diagnostic/Feasibility Study; Restoration/Protection Implementation Program; and Post-Restoration Monitoring.

- **Section 319, Nonpoint Source (NPS) Program Implementation** is the successor to the Section 208 program conducted in the late 1970s and early 1980s. Only those activities that implement the state's EPA-approved Nonpoint Source Management Plan are eligible for funding. This includes basic program implementation tasks, provided these activities help to institutionalize the program within a state, watershed NPS management program implementation within targeted watersheds, NPS control practice installation at demonstration sites, and groundwater NPS assessment and management programs.
- **Section 320, National Estuary Program** authorizes the development of comprehensive conservation and management plans, usually over a five-year period, for specific legislatively designated estuaries. Does not provide funding for the implementation of approved plans although other CWA funds (i.e., 319, Title II and VI) may be used.
- **Section 401, Water Quality Certification** requires that any applicant for a federal permit to conduct any activity, including the construction or operation or a facility which may result in the discharge of any pollutant, must obtain certification of those activities from the state in which the discharge originates.
- **Section 402, National Pollutant Discharge Elimination System (NPDES),** established the NPDES to regulate the discharge of pollutants from point sources. The NPDES permit program is administered by the EPA and the responsibility for administration and enforcement of the program at the state level has been delegated to certain agencies in states that have primacy.
- **Section 402(p) of the Clean Water Act** has among its requirements that EPA established regulations setting forth NPDES permit applications for stormwater discharges from industrial sources and municipal separate storm sewer (drain) systems. EPA published draft regulations addressing stormwater discharges on December 17, 1990. The regulations are administered through the existing NPDES program. The rule sets forth the required components of municipal stormwater management plans and specifies who is covered by the regulations, the nature of the requirements, and a one-to-two-year time schedule for compliance with the permit application requirements. The regulations also set forth application requirements for industries. Included are requirements for a stormwater permit for all construction activities that disturb an area of 2 ha (5 ac) or greater. Some states have NPDES permitting authority. For those who do not, authority is through the EPA regional offices. A more detailed summary of the NPDES program is provided following.
- **Section 404, Permits for Dredged or Fill Material** allows establishment of a regulatory program to control the discharge of dredged or fill material into navigable waters (wetlands). This program is administered by the U.S. Army Corps of Engineers, using permitting guidelines developed in coordination with EPA, although the section allows a state to assume administration of the program
- **Section 604, Title VI Water Quality Management Planning** requires each state to reserve one percent of the State Revolving Loan Fund grant for water quality management planning activities required by Section 205(j) and 303(e). Eligible activities include projects to determine the nature, extent and causes of water quality problems; to identify cost-effective and acceptable point and nonpoint source controls; and to develop implementation plans.

B.1.2 NPDES Program (Phases I and II). In 1990, EPA published its Final Rule for regulation of stormwater discharges. This rule (40 CFR 121.26), known as the Phase I Rule, established requirements for the permitting and management of stormwater discharges from large (serving a population of 250,000 or more) and medium-sized (serving a population of 100,000 to 250,000) municipalities, numerous industries (based on SIC code), and construction sites disturbing 2 ha (5 ac) or more. The permits are administered by the EPA (or certain delegated state agencies) under the NPDES program. Recent additions to NPDES-related guidance documents can be accessed online at http://cfpub.epa.gov/npdes.

Phase I Municipal Requirements. Under the Phase I Rule, coverage under an EPA Construction General Permit must be sought by the operator of a construction activity that

- Will disturb 2 ha (5 ac) or greater, or
- Will disturb less than 0.40 ha (1 ac) but is part of a larger common plan of development or scale whose total land-disturbing activities total one acre or greater (or is designated by the NPDES permitting authority); and
- Will discharge stormwater runoff from the construction site into a municipal separate storm sewer system (MS4) or waters of the United States.

In December 1999, EPA finalized the Phase II regulations, which require controls on stormwater discharges from a broader sector of municipalities, industries, and construction sites. Specifically, for construction (USEPA 1999), the Phase II Rule requires construction sites disturbing equal to or greater than 0.40 ha (1 ac) and less than 2 ha (5 ac) to control pollutants in stormwater runoff. Construction activity disturbing less than 0.40 ha (1 ac) requires a permit if it is part of a larger common plan of development or sale disturbing a total of 0.40 ha (1 ac) or greater or is individually designated for permit coverage by the NPDES permitting authority.

Nationally, the Phase II Rule requires operators of Phase II construction sites to obtain an NPDES permit and implement BMPs to minimize pollutant runoff. For the Phase II construction program, EPA has taken an approach similar to Phase I where the program requirements are not fully defined in the rule but rather in the NPDES stormwater permit issued by the NPDES permitting authority. The Phase II requirements are similar to the following three main requirements of EPA's Construction General Permit for Phase I:

- Submittal of a NOI that includes general operator and site information
- The development and implementation of a SWPPP with appropriate BMPs to minimize the discharge of pollutants from the site
- Submittal of a NOT when final stabilization of the site has been achieved as defined in the permit, or stormwater runoff is no longer being discharged, or when another operator has assumed control of the site.

A copy of EPA's overview of the Storm Water Phase II Final Rule for Construction Sites can be obtained from the EPA (USEPA 1999).

Phase II Municipal Requirements. Phase II municipalities must apply for an NPDES stormwater permit, develop a Storm Water Management Plan (SWMP) that includes the six minimum control measures, implement the SWMP using appropriate BMPs, develop measurable program goals, and evaluate program effectiveness. The applicable standards for Phase II municipalities are to reduce the discharge of pollutants to the "maximum extent practicable" (MEP), protect water quality, and satisfy the water quality requirements of the Clean Water Act.

The six minimum control measures are

1. Public education and outreach
2. Public participation and involvement
3. Illicit discharge detection and elimination
4. Construction site runoff control
5. Postconstruction runoff control
6. Pollutant prevention/good housekeeping.

It should be noted that control measures 4, 5, and 6 apply directly to construction site ESC programs. The other three control measures are more programmatic.

Minimum Measure 1–Public Education and Outreach. This measure includes distributing educational materials and performing outreach to inform citizens about the impacts that polluted stormwater runoff discharges can have on water quality.

Minimum Measure 2–Public Participation and Involvement. This measure includes providing for citizens to participate in program development and implementation, including effectively publicizing public hearings and encouraging citizen representatives to participate on a stormwater management panel.

Minimum Measure 3–Illicit Discharge Detection and Elimination. This measure includes developing and implementing a plan to detect and eliminate illicit discharges to the storm drain system (including developing a system map and informing the community about hazards associated with illegal discharges and improper waste disposal).

Minimum Measure 4–Construction Site Runoff Control. This measure includes developing, implementing, and enforcing an erosion and sediment control program for construction activities that disturb 0.40 ha or more (1 or more acres) of land.

Minimum Measure 5–Postconstruction Runoff Control. This measure includes developing, implementing, and enforcing a program to address discharges of postconstruction stormwater runoff from new development and redevelopment areas.

Minimum Measure 6–Pollution Prevention/Good Housekeeping. This measure includes developing and implementing a program with the goal of preventing or reducing pollutant runoff from municipal operations. The program must include municipal staff training on pollution prevention measures (e.g., street sweeping, catch basin cleaning, pesticide use reduction).

B.1.3 State Regulations Run by EPA. The EPA is responsible for implementing the stormwater NPDES program. In states that have NPDES delegated authority, this process is managed by the state environmental agency.

Types of Stormwater Permits. The EPA (and certain states that have NPDES permitting authority) has developed NPDES construction permits for construction sites disturbing 0.40 ha or more (1 acre or more). The party responsible for compliance with the general permit is the owner of the land on which the construction is to occur. There are two categories of stormwater NPDES permits: individual permits and general permits. The vast majority of stormwater dischargers are covered under general permits; individual permits are rare. All general permits cover only stormwater discharges, and all require the preparation of a SWPPP.

Individual permits are issued to a specific construction project at a specific location. Individual permits are required for discharges that are already covered under an individual permit and for discharges contributing to a violation of water quality standards.

Types of Permitting Authority. NPDES stormwater permits are issued by the EPA and certain delegated state agencies. Almost all states have general permitting authority. Only some states and territories have no NPDES authority at all.

Permit Application Requirements. The permit application process for obtaining coverage under a general permit involves filing a NOI. Obtaining an individual permit is more complex and requires the development of an individual permit application, which includes extensive information about the site.

State-Specific Requirements. State-specific requirements for general permits are contained in the permits themselves and may be obtained from each state. EPA general permits may be obtained from EPA regional offices.

Notice of Intent for Construction Sites. The NOI informs regulators that the facility or construction project will be in compliance with a general permit. The NOI form does not require any detailed information. The specific format and information requirements vary among EPA and the states. In general, the construction site NOI requires the following information:

- Brief project description
- Estimated timetable for major activities
- Estimate of area to be disturbed
- Certification that the SWPPP will provide compliance with state and local requirements (including the signature of a responsible party).

Notice of Termination for Construction Sites. A NOT must be submitted to the permitting authority in order to terminate coverage. A permittee may submit a NOT when

- Disturbed soils at the construction site have been stabilized and temporary erosion and sediment control measures have been removed (or will be removed at an appropriate time)
- Stormwater discharges have been eliminated
- The permittee is no longer an operator of the site.

B.1.4 NPDES General Construction Stormwater Permits

Storm Water Pollution Prevention Plan Requirements. The permit requires the preparation of a SWPPP. One key element of the SWPPP is the requirement for preparation of an erosion and sediment control plan.

Specifically, the SWPPP must include

- A topographic map
- Site map
- Narrative description
- List of potential pollutants
- Estimate of size of the site
- Estimate of the runoff coefficient before and after construction
- Estimate of the percent imperviousness of the site before and after construction
- A copy of the NOI
- An erosion and sediment control plan

- Nonstormwater management control measures
- Postconstruction stormwater management controls
- Waste management and disposal measures
- Provisions for maintenance, inspection, and repair according to BMPs
- Provisions for training
- A subcontractors/contractors list.

In addition, the SWPPP must include information on

- Public access to the SWPPP
- Other relevant plans
- The preparer of the plan
- Site inspections before, during, and after storms
- Implementation documentation
- Noncompliance reporting.

In general, the requirements for erosion and sediment control under the NPDES construction permits are

- Soil stabilization practices designed to preserve existing vegetation and revegetate open areas after grading
- Structural practices to divert flows from exposed areas, store flows, or otherwise limit runoff from exposed areas
- Control practices to reduce tracking of sediment and wind erosion.

Deadlines. The Phase I regulations for construction site disturbing 2 ha (5 ac) or more are currently in place. Applications for construction general permits must be filed at least 2 days prior to the start of construction. In addition, the SWPPP must be prepared before the NOI is filed.

The Phase II regulations for construction sites disturbing 0.40 ha (1 ac) or more but less than 2 ha (5 ac) became final on December 8, 1999. Under the Phase II Rule, general permits for small construction sites must be developed by December 8, 2002. Small construction sites must obtain coverage under a construction general permit by March 10, 2003.

Certification and Notification. The SWPPP must be signed by an authorized representative of the site operator. This certifies that the information is true and assumes liability for contractors. Contractors also must sign the certification.

Enforcement and Penalties. EPA has three types of enforcement options:

- Administrative orders
- Civil actions
- Criminal prosecutions.

Under administrative order, EPA can impose fines and penalties without court action. The maximum fine is more than $15,000 per day, with a maximum of more than $150,000. The fine must consider the nature of the violation, circumstances, prior history, and other factors.

In a civil action, EPA may bring suit without an administrative order. The maximum penalty is more than $35,000 per day with no maximum limit. The court must consider the seriousness of the violation, past history of violations, and other factors.

Finally, EPA may refer the case to the Department of Justice for criminal prosecution. Penalties may range from $50,000 per day to more than $1,000,000 and may include imprisonment. Violations may include failure to maintain proper records, failure to implement BMPs, and other infractions.

NPDES-delegated states have the authority to enforce Clean Water Act violations. If the EPA finds a violation of the Clean Water Act, it must pursue the violation or notify the state. If the state does not take action, then EPA must enforce. States have access to all enforcement options under the Clean Water Act (administrative orders, civil actions, and criminal prosecutions). In addition, many states have more stringent penalties than the EPA.

The Clean Water Act allows formal challenges to be made by third parties regarding compliance with the regulations. These challenges may be brought by parties adversely affected by the selected stormwater quality controls, state or federal water quality regulatory agencies, national-scale environmental activist organizations, local citizens' groups, or environmental organizations.

The basis for these formal challenges includes (1) the planning and decision making process was "arbitrary and capricious," (2) the process failed to adequately consider relevant information, or (3) the process failed to involve knowledgeable people in a timely manner.

B.1.5 Federal Safe Drinking Water Act

Section 1442, Wellhead Protection provides technical assistance and funding to states and local governments designing and implementing wellhead protection programs (WHP). Eligible activities include delineation of WHP areas; identification, mapping, and sampling of contamination sources; public education; and development of ordinances.

Section 1443, Underground Injection Control establishes federal and state programs to protect groundwater from these sources. It provides grants to states to fund all types of activities related to this program.

B.2 U.S. DEPARTMENT OF AGRICULTURE

B.2.1 Environmental Quality Incentives Program (EQIP). This section provides technical, educational, and financial assistance to eligible farmers to address soil, water, and related natural resource concerns on their lands in an environmentally beneficial and cost-effective manner. The program is funded through the Commodity Credit Corporation. The purposes of the program are achieved through the implementation of a conservation plan. Five-to-10-year contracts are made with producers, and cost-share payments may be made to implement eligible conservation practices including animal waste management systems, buffer strips, and nutrient management.

B.2.2 Conservation Reserve Program (CRP). The Conservation Reserve Program is intended to return certain agricultural lands that are highly erodible or otherwise critical in protecting and restoring water quality to a conservation use, typically as forests. Farmers receive payment on a per acreage basis to conserve these lands for a contracted period, typically 10 years.

B.2.3 Wetlands Reserve Program (WRP). The Water Reserve Program is intended to restore and protect farmed wetlands or converted wetlands. Farmers receive direct payments as well as conservation planning and technical assistance to install necessary restoration practices on those areas that they agree to maintain under a conservation easement.

B.2.4 Resource Conservation and Development (RC&D) Program. The Resource Conservation and Development Program is administered by the NRCS (formerly the Soil Conservation Service), which encourages and improves the capability of state and local entities in rural areas to plan, develop, and implement programs, typically in targeted critical areas.

B.2.5 Soil and Water Conservation. The Soil and Water Conservation program is administered by the NRCS, provides technical assistance to the public through total resource planning and management to improve water quality, natural resources, and reduce pollution sources. NRCS also has been publishing detailed soil surveys for each county in the country providing a wide variety of useful natural resources management information.

B.2.6 Watershed Protection and Flood Prevention (Small Watershed Program; PL-566 Program). This program, administered by the NRCS, provides technical and financial assistance to state agencies and local governments in the development and implementation of plans to protect, develop, and use land and water resources in small watersheds. Recently, this program has become more broadly oriented with greater emphasis on protecting and restoring water quality, especially from problems related to flooding, erosion, sedimentation, and use or disposal of water.

B.3 U.S. DEPARTMENT OF THE INTERIOR

B.3.1 National Water Quality Assessment Program. This program, administered by the U.S. Geological Survey (USGS), addresses a wide range of major water quality issues, with special emphasis on pesticide impacts on water resources. The program includes nationwide surface and groundwater quality monitoring and assessment.

B.3.2 Water Data Program. Also administered by the USGS, this program consists of four water quality monitoring networks, the most important of which is the National Stream Quality Accounting Network. Data on stream flow and height, lake stage and storage, groundwater levels, well and spring discharge, and the quality of surface and groundwater are collected and stored.

As part of USGS's program of releasing water data to the public, a large-scale computerized system has been developed for the storage and retrieval of water data collected through its activities. The National Water Data Storage and Retrieval System (WATSTORE) was established in 1972 to provide an effective and efficient means for the processing and maintenance of water data collected through the activities of the USGS and to facilitate release of data to the public. A variety of useful products, ranging from data tables to complex statistical analyses (such as log-Pearson Type III), can be produced using WATSTORE. The system resides on the central computer facilities of the USGS at its National Center in Reston, Virginia, and consists of related files and databases (http://waterdata.usgs.gov/nwis).

B.3.3 Federal-State Cooperative Program This program establishes a partnership for water resources investigations between the USGS and state and local agencies. This program is the foundation for much of the planning, development, and management of the nation's water resources.

B.3.4 Coastal Wetlands Planning, Protection and Restoration Program. This program, administered by the Fish and Wildlife Service, provides funds for the acquisition of coastal lands or waters and for restoration, enhancement, or management of coastal wetland ecosystems. Projects must provide for the long-term conservation of these lands and waters.

B.3.5 Land and Water Conservation Program. This program, administered by the National Park System, was established to create and maintain a national legacy of high-quality recreation areas. The program provides funding for federal acquisition of authorized national park, conservation, and recreation areas, and to state and local governments to help them acquire, develop, and improve outdoor recreation areas.

B.4 FEDERAL HIGHWAY ADMINISTRATION

The Federal Aid Highway Program assists state agencies in the development and improvement of an integrated, interconnected transportation system. Funds may be used for planning, research, and development (including BMPs), restoration, roadside beautification, and wetland mitigation. This program provides funding for erosion and sediment controls needed to minimize highway construction impacts but not typically for the treatment and management of highway runoff. However, the Surface Transportation Program component supposedly authorizes funding for highway stormwater quality controls and for mitigating damage to ecosystems, habitat, and wildlife.

B.5 U.S. ARMY CORPS OF ENGINEERS

Through its Civil Works Program, Congress has authorized the U.S. Army Corps of Engineers (Corps) to conduct planning, design, and construction for water resources projects for navigation, flood damage reduction, and ecosystem restoration, as well as storm damage reduction, hydroelectric power, environmental infrastructure, recreation, and water supply. The Corps has two major categories of civil works projects. *Specifically authorized projects* receive specific authorizations from Congress for the start of planning and again prior to initiation of construction. The Corps also has nine *continuing authorities* for water resources projects under its Continuing Authorities Program.

There are numerous specifically authorized projects across the nation to address erosion and sedimentation issues at the local level and as regional watershed studies. Under Section 729 of the Water Resources Development Act of 1986, as amended, Congress has provided the Corps with authority to conduct watershed studies within the United States. These watershed studies can investigate all aspects of watershed water resource planning, including problems and solutions related to erosion and sedimentation.

The Corps' Continuing Authorities Program includes authorities for ecosystem restoration and for flood damage reduction. The ecosystem restoration authorities under the Continuing Authorities Program are Section 206 of the Water Resources Development Act of 1996, as amended, and Section 1135 of Water Resources Development Act of 1986, as amended. Section 206 allows the Corps to plan, design, and construct projects for aquatic ecosystem restoration and protection to improve environmental quality. Section 1135 provides the Corps with authority to modify existing Corps projects to restore degradation to ecosystems resulting from Corps projects. The Continuing Authorities Program also includes an authority that permits the Corps to make beneficial use of dredged material for aquatic habitat restoration under Section 204 of the Water Resources Development Act of 1992, as amended.

The flood reduction authorities included under the Continuing Authorities Program include Section 205 of the Flood Control Act of 1948, as amended; Section 208 of the Flood Control Act of 1954, as amended; Section 14 of the Flood Control Act of 1946, as amended; and Section 103 of the Rivers and Harbors Act of 1962, as amended. Section 205 permits the Corps to plan, design, and construct small flood damage reduction projects. Section 205 projects can include multiple purposes, including ecosystem restoration. Section 208 permits the Corps to study, design, and construct in-stream clearing and snagging projects to prevent local flooding. Section 14 allows the Corps to study, design, and construct emergency streambank protection measures to protect public highways, utilities, and other public facilities. Under Section 103, the Corps is authorized to study, design, and construct beach erosion protection projects to reduce damage from storm-induced waves and currents.

Under its program for Planning Assistance for the States (Section 22 of the Water Resources Development Act of 1974, as amended), the Corps provides technical assistance to states, the District of Columbia, Puerto Rico, the Virgin Islands, Guam, American Samoa, Commonwealth of the Mariana Islands, Palau, and U.S. tribal nations to prepare comprehensive water and related land resources development plans, including watershed and ecosystem studies that may include investigations of sedimentation and erosion. The Tribal Partnership Program under Section 203 of the Water Resources Development Act of 2000 allows the Corps to conduct studies for water and related land resources, including flood damage reduction, ecosystem restoration and protection, and preservation of cultural and natural resources, to benefit Indian tribes.

The Corps of Engineers also has been provided with general authority to provide flood plain information and planning assistance to state, county, and city governments, and to other federal agencies. Public Law 84-99 allows the Corps to participate in planning and preparedness for natural disasters, in emergency repair and restoration of flood or shore protection projects damaged by storms or floods, and in advance measures to prevent loss of life and catastrophic property damage when there is an imminent threat of unusual flooding.

Under its Regulatory Program, the Corps of Engineers and the Secretary of the Army must approve plans for the construction of any dam or dike across any navigable water of the United States. The Corps also issues permits for discharges of dredged or fill materials into the waters of the United States under Section 404 of the Clean Water Act, and issues permits for the transportation of dredged materials for ocean dumping under Section 103 of the Marine Protection, Research and Sanctuaries Act of 1972.

B.6 NATIONAL OCEANIC AND ATMOSPHERIC ADMINISTRATION

B.6.1 Coastal Zone Management (CZM) Act of 1972. This act allows states to prepare and implement comprehensive management programs for coastal resources, which balance competing demands on resource protection, protection of public health and safety, provision for public access, and economic development. States with federally approved programs receive federal grants to develop and implement a wide variety of coastal resource management initiatives.

B.6.2 Section 6217, Coastal Nonpoint Pollution Control Program. This program is jointly administered with the EPA, which is responsible for establishing minimum NPS management measures. States with federally approved CZM programs must develop and implement programs to restore and protect coastal waters, which include compliance with the minimum NPS management measures.

B.6.3 National Estuarine Research Reserve System. This system allows establishment and management of a national system of reserves representing different coastal regions and estuarine types. The reserves serve as field laboratories and as public education centers.

B.6.4 National Marine Sanctuary Program. This program allows identification of areas of the marine environment of special significance and provides authority for comprehensive and coordinated conservation and management of these areas. This program also provides for research and monitoring activities and for public education.

APPENDIX C

STATE EROSION AND SEDIMENT CONTROL REGULATORY AGENCIES WITHIN THE UNITED STATES

Alabama
Soil and Water Conservation Committee
P.O. Box 304800
Montgomery, AL 36130-4800

Alaska
Department of Environmental Conservation
410 Willoughby Ave., Suite 303
P.O. Box 111800
Juneau, AK 99811-1800

Arizona
Department of Environmental Quality
1110 W. Washington St.
Phoenix, AZ 85007

Arkansas
State Highway and Transportation Department
10324 Interstate 30
Little Rock, AK 72209

California
CA Environmental Protection Agency
1001 "I" St.
P.O. Box 2815
Sacramento, CA 95812-2815

Colorado
Water Quality Control Division
4300 Cherry Creek Drive South
Denver, CO 80246-1530

Connecticut
Department of Environmental Protection
79 Elm St.
Hartford, CT 06106-5127

Delaware
Delaware Sediment & Stormwater Program
DNREC Division of Watershed Stewardship
89 Kings Hwy.
Dover, DE 19901

Florida
Nonpoint Source Management Program
2600 Blair Stone Road
Mail Station 3570
Tallahassee, FL, 32399-2400

Georgia
Georgia Environmental Protection Division
Georgia Department of Natural Resources
2 Martin Luther King Jr. Drive, Suite 1152 East Tower
Atlanta, GA 30334

Hawaii
Clean Water Branch, Environmental Management Division
State Department of Health
P.O. Box 3378
Honolulu, HI 96801-3378

Idaho
Idaho Department of Environmental Quality
Water Quality Division
1410 N. Hilton St.
Boise, ID 83706

Illinois
Illinois EPA
1021 North Grand Ave. East
P.O. Box 19276
Springfield, IL 62794-9276

Indiana
Office of Water Quality
Indiana Department of Environmental Management
Government Center North
100 N. Senate Ave.
Indianapolis, IN 46204

Iowa
Water Quality Bureau, Department of Natural Resources
502 East 9th St.
Des Moines, IA 50319-0034

Kansas
Bureau of Water, Industrial Programs Section
1000 SW Jackson St., Suite 420
Topeka, KS 66612-1367

Kentucky
Kentucky Division of Water
200 Fair Oaks Lane, Fourth Floor
Frankfort, KY 40601

Louisiana
Department of Environmental Quality
602 North Fifth St.
Baton Rouge, LA 70802

Maine
Department of Environmental Protection
17 State House Station
Augusta, ME 04333-0017

Maryland
Maryland Department of the Environment
1800 Washington Blvd.
Baltimore, MD 21230

Massachusetts
Department of Environmental Protection
One Winter Lane
Boston, MA 02108

Michigan
Department of Environmental Quality
Constitution Hall, Fifth Floor
505 West Allegan St.
Lansing, MI 48933

Minnesota
Pollution Control Agency, Storm Water Program
520 Lafayette Road North
St. Paul, MN 55155-4194

Mississippi
Department of Environmental Quality
P.O. Box 2261
Jackson, MS 39225

Missouri
Missouri Department of Natural Resources
P.O. Box 176
Jefferson City, MO 65102

Montana
Department of Environmental Quality
Lee Metcalf Building
1520 East Sixth Ave.
P.O. Box 200901
Helena, MT 59620-0901

Nebraska
Nebraska Department of Environmental Quality
1200 "N" St., Suite 400
P.O. Box 98922
Lincoln, NE 68509

Nevada
Division of Environmental Protection
901 South Stewart St., Suite 4001
Carson City, NV 89701–5249

New Hampshire
Department of Environmental Services
29 Hazen Drive
P.O. Box 95
Concord, NH 03302-0095

New Jersey
Division of Water Quality
Department of Environmental Protection
P.O. Box 402
Trenton, NJ 08625-0402

New Mexico
New Mexico Environment Department
Harold L. Runnels Building
1190 St. Francis Drive, Suite N4050
Santa Fe, NM 87505

New York
Department of Environmental Conservation
625 Broadway
Albany, NY 12233-0001

North Carolina
Land and Water Stewardship
NC Department of Environmental Quality
1601 Mail Service Center
Raleigh, NC 27699-1601

North Dakota
Division of Water Quality
918 East Divide Ave., Fourth Floor
Bismarck, ND 58501-1947

Ohio
Division of Soil and Water Resources
Building B-3
2045 Morse Road
Columbus, OH 43229

Oklahoma
Oklahoma Conservation Commission
2800 N. Lincoln Blvd., Suite 160
Oklahoma City, OK 73105

Oregon
Department of Environmental Quality
811 Sixth Ave.
Portland, OR 97204-1390

Pennsylvania
Department of Environmental Protection
Rachel Carson State Office Building
400 Market St.
Harrisburg, PA 17101

Rhode Island
Department of Environmental Management
235 Promenade St.
Providence, RI 02908-5767

South Carolina
Department of Health and Environmental Control
2600 Bull St.
Columbia, SC 29201

South Dakota
Department of Environmental and Natural Resources
Joe Foss Building
523 East Capitol St.
Pierre, SD 57501

Tennessee
Department of Environment and Conservation
L&C Tower
401 Church St.
Nashville, TN 37243

Texas
Commission on Environmental Quality
12100 Park 35 Circle
Austin, TX 78753

Utah
Division of Water Quality
DEQ, Third Floor
195 North, 1950 West
P.O. Box 144870
Salt Lake City, UT 84116

Vermont
Agency of Natural Resources
Center Building
103 South Main St.
Waterbury, VT 05671-0301

Virginia
Department of Conservation and Recreation
Soil and Water Conservation Programs
203 Governor St., Suite 206
Richmond, VA 23219-2094

Washington
Department of Ecology
P.O. Box 47600
Olympia, WA 98504-7600

West Virginia
Department of Environmental Protection
601 57th St., S.E.
Charleston, WV 25304

Wisconsin
Department of Natural Resources
101 South Webster St.
P.O. Box 7921
Madison, WI 53707-7921

Wyoming
Department of Environmental Quality
122 West 25th St., 4-W
Cheyenne, WY 82002

APPENDIX D
GLOSSARY

Note: Items in this glossary marked with an asterisk (*) are from District of Columbia (2003).

Aesthetic—Pleasing in appearance; showing good taste.

Antiseep Collar*—An impermeable diaphragm usually of sheet metal or concrete constructed at intervals within the zone of saturation along the conduit of a principal spillway to increase the seepage length along the conduit and thereby prevent piping or seepage along the conduit.

Apron*—A floor or lining to protect a surface from erosion, for example, the pavement below chutes, spillways, or at the toes of dams.

Baffles*—Vanes, guides, grids, grating, or similar devices placed in a conduit to deflect or regulate flow and effect a more uniform distribution of velocities.

Barrel*—Usually mild sloping closed conduit used to convey water under or through a dam; part of a principal spillway.

Bedload*—The sediment that moves by sliding, rolling, or bounding on or very near the streambed; sediment moved mainly by tractive or gravitational forces or both but at velocities less than the surrounding flow.

Beneficial Uses—Uses of water necessary for the survival and well-being of people, plants, and wildlife. Examples of beneficial uses include use as drinking water, swimming, fishing, industrial and agricultural supply, and support of habitat.

Berm*—A shelf or small dike that breaks the continuity of a slope.

Best Management Practices (BMPs)—Techniques, measures, or structural controls used to manage the quantity and improve the quality of stormwater runoff.

Biodegradable*—Material capable of being broken down (degraded) by common soil organisms.

Channel Stabilization*—Erosion prevention and stabilization of velocity distribution in a channel using jetties, drops, revetments, structural linings, vegetation, and other measures.

Check Dam*—A small dam constructed in a gully or other small watercourse to decrease the stream flow velocity (by reducing the channel gradient), minimize channel scour, and promote deposition of sediment.

Clay (Soils)*—1. A mineral soil separate consisting of particles less than 0.002 mm (0.00008 in.) in equivalent diameter. 2. A soil texture class. 3. (Engineering). A fine-grained soil (more than 50% passing the No. 200 sieve) that has a high plasticity index in relation to the liquid limit (see Unified Soil Classification System).

Compaction*—To unite firmly. With respect to construction work with soils, engineering compaction is any process by which the soil grains are rearranged to decrease void space and bring them into closer contact with each other, thereby increasing the weight of solid material per unit of volume, increasing the shear and bearing strength and reducing permeability.

Conduit*—Any channel intended for the conveyance of water, whether open or closed.

Construction Site Operator—According to EPA's General Construction Permits, a construction site operator is someone who maintains overall operational control over construction plans and specifications, including the ability to change those plans. Typically, this refers to the developer of a construction site. Another defining characteristic can be maintaining day-to-day operational control over activities that will ensure compliance with the SWPPP. Typically, this refers to the general or subcontractor.

Contour*—1. An imaginary line on the surface of the earth connecting points of the same elevation. 2. A line drawn on a map connecting points of the same elevation.

Crest*—1. The top of a dam, dike, spillway, or weir, frequently restricted to the overflow portion. 2. The summit of a wave or peak of a flood.

Critical Area or Site*—Sediment-producing, highly erodible or severely eroded areas.

Crushed Stone*—Aggregate consisting of angular particles produced by mechanically crushing rock.

Cut*—Portion of land surface or area from which earth has been removed or will be removed by excavation; the depth below original ground surface to excavated surface.

Cut-and-Fill*—Process of earth moving by excavating part of an area and using the excavated material for adjacent embankments or fill areas.

CZARA—Section 6217 of the Coastal Zone Reauthorization Amendments of 1990 (CZARA) requires the development of a Coastal Nonpoint Pollution Control Program Management Plan for every designated state under coastal zone jurisdiction. The purpose of each plan is to address the nonpoint source pollution concerns for coastal zone areas.

Dam*—A barrier to confine or raise water for storage or diversion, to create a hydraulic head, to prevent gully erosion, or for retention of soil, sediment, or other debris.

Debris*—Remains of plants, objects, and rocks that form trash.

Deposition*—The accumulation of material dropped because of a slackening movement of the transporting agent, water, or wind.

Dike (Engineering)*—An embankment to confine or control water, for example, one built along the banks of a river to prevent overflow or lowlands; a levee.

Disturbed Area*—An area in which the natural vegetative soil cover has been removed or altered and, therefore, is susceptible to erosion.

Diversion*—A channel with a supporting ridge on the lower side constructed across the slope to divert water from areas where it is in excess to sites where it can be used or disposed of safely. Diversions differ from terraces in that they are individually designed.

Drain (Noun)*—1. A buried pipe or other conduit (subsurface drain). 2. A ditch or channel (open drain) for carrying surplus surface water or groundwater.

Drain (Verb)*—1. To provide channels, such as open ditches or closed drains, so that excess water can be removed by surface flow or internal flow. 2. To lose water (from the soil) by percolation.

Drainage*—1. The removal of excess surface water or groundwater from land by means of surface or subsurface drains. 2. Soils characteristics that affect natural drainage.

Drainage Area—All land and water area from which runoff may run to a common (design) point.

Drop Inlet—An outfall structure in which the water drops through a vertical riser connected to a discharge conduit.

Drop Structure*—A structure for dropping water to a lower level and dissipating surplus energy; a fall. The drop may be vertical or inclined.

Emergency Spillway*—A dam spillway designed and constructed to discharge flow in excess of the principal spillway design discharge.

Erosion*—1. The wearing away of the land surface by running water, wind, ice, or other geological agents, including such processes as gravitational creep. 2. Detachment and movement of soil or rock fragments by water, wind, ice, or gravity. The following terms are used to describe different types of water erosion:

- *Accelerated erosion*—Erosion much more rapid than normal, natural or geological erosion, primarily as a result of the influence of the activities of humans or, in some cases, of animals or natural catastrophes that expose base surfaces, for example, fires.
- *Gully erosion*—The erosion process whereby water accumulates in narrow channels and, over short periods, removes the soil from this narrow area to considerable depths.
- *Rill erosion*—An erosion process in which numerous small channels only a dozen or so millimeters deep are formed (see Rill).
- *Sheet erosion*—The spattering of small soil particles caused by the impact of raindrops on wet soils. The loosened and spattered particles may or may not subsequently be removed by surface runoff.

Erosive Velocities*—Velocities of water that are high enough to wear away the land surface. Exposed soil will generally erode faster than stabilized soils. Erosive velocities vary according to the soil type, slope, and structural or vegetative stabilization used to protect the soil.

Fines (Soil)*—Generally refers to the silt and clay size particles in soil.

Gabion*—A flexible woven-wire basket composed of two to six rectangular cells filled with small stones. Gabions may be assembled into many types of structures such as revetments, retaining walls, channel liners, drop structures, and groins.

Grade*—1. The slope of a road, channel, or natural ground. 2. The finished surface of a canal bed, road bed, top of embankment, or bottom of excavation; any surface prepared for the support of construction, such as paving or laying a conduit. 3. To finish the surface of a canal bed, road bed, top of embankment or bottom of excavation.

Grassed Waterway*—A natural or constructed waterway, usually broad and shallow, covered with erosion-resistant grasses, to convey surface water down the slope.

Gravel*—1. Aggregate consisting of mixed sizes of particles 0.25 in. to 3 in. (6.35 mm to 76.2 mm) that normally occur in or near old streambeds and have been worn smooth by the action of water. 2. A soil having particle sizes, according to the Unified Soil Classification System, with greater than 50% retained by No. 4 sieve size and angular in shape as produced by mechanical crushing.

Ground Cover*—Plants that are low-growing and provide a thick growth, which protects the soil as well as providing some beautification of the area occupied.

Gully*—A channel or miniature valley cut by concentrated runoff through which water commonly flows only during and immediately after heavy rains or during the melting of snow. The distinction between a gully and a rill is one of depth. A gully is sufficiently deep that it would not be obliterated by normal tillage operations, whereas a rill is shallower depth and would be smoothed by ordinary farm tillage.

Herbicide*—Chemical formulation used to control weeds or brush.

Inflow Protection*—A water-handling device used to protect the transition area between any water conveyance (dike, swale, or swale dike) and a sediment-trapping device.

Larger Common Plan of Development or Sale—EPA's construction general permits define a larger common plan of sale as "a contiguous area where multiple separate and distinct construction activities are occurring under one plan." The "plan" is broadly defined as any announcement or piece of documentation or physical demarcation indicating construction activities may occur on a specific plot.

For example, a developer plans to clear and grade a 4 ha (10-acre) site for a new subdivision of single-family homes. The developer and contractor obtain permit coverage for their activities. Upon clearing and grading the site, the developer sells 20 lots of a 0.20 ha (0.5 ac) each to individual homebuilders. Because he or she is no longer responsible for the site (i.e., all interest has been sold in the site), the developer can terminate coverage. The new operator of each 0.20 ha (0.5 ac) site must then obtain permit coverage, even though the operator will not disturb greater than 2.0 ha (5 ac). This requirement remains in effect regardless of any lapse in time between the initial grading or clearing and the actual construction of the home.

Lime*—Basic calcareous materials used to raise pH of acid soils for benefit of plants being grown. May be either ground limestone or hydrated lime.

Mulch*—Covering on the surface of soil to protect and enhance certain characteristics, such as water retention qualities.

Netting (Mulch)*—Paper or cotton material used to hold mulch material on the soil surface.

Nonerosive Velocity*—Controlling the velocity of water to prevent detachment and movement of soil or rock. Erosive velocity varies according to the soil type, slope, structural, or vegetative stabilization used to protect the soil.

Outfall*—The point where water flows from a conduit, stream, or drain.

Outlet*—The point at which water discharges from such things as a stream, river, lake, tidal basin, pipe, channel, or drainage area.

Outlet Channel*—A waterway constructed or altered primarily to carry water from artificially constructed structures, such as terraces, subsurface drains, diversions, and impoundments.

Parent Material*—Unconsolidated rock material from which the soil profile develops.

Permanent Seeding*—Results in establishing perennial vegetation that may remain on the area for many years.

Permissible Velocity (Hydraulics)*—1. The highest average velocity at which water may be carried safely in a channel or other conduit. 2. The highest velocity that can exist through a substantial length of a conduit and not cause scour of the channel. 3. A safe, non-eroding or allowable velocity.

Piping*—Removal of soil material through subsurface flow channels or "pipes" developed by seepage water.

Recycled Concrete Equivalent*—May be substituted for stone or riprap as called for in specifications except where stone must be "washed." Concrete used in place of stone shall be broken into the stone sizes required for the application, shall be angular, resist crumbling, and shall contain no steel reinforcement.

Right-of-Way*—1. Right of passage, as over another's property. 2. A route that is lawful to use. 3. A strip of land acquired for transport or utility construction.

Rill*—A small channel cut by concentrated runoff but through which water commonly flows only during and immediately after rains or during the melting of snow. A rill is usually only a dozen or so millimeters deep (but no more than a third of a meter) and, hence, no obstacle to tillage operations.

Riprap*—Broken rock, cobbles, or boulders placed on earth surfaces, such as the face of a dam or the bank of a stream, for protection against the action of water (waves); also applies to brush or pole mattresses, or brush and stone, or similar materials used for soil erosion control.

Runoff (Hydraulics)*—That portion of the precipitation on a drainage area that is discharged from the area in the stream channels. Types include surface runoff, groundwater runoff or seepage.

Sand*—1. (Agronomy) A soil particle between 0.05 and 2.0 mm (0.002 in. and 0.08 in.) in diameter. 2. A soil textural class. 3. (Engineering) According to the Unified Soil Classification System, a soil particle larger than the No. 200 sieve (0.074 mm [0.0029 in.]) and passing the No. 4 sieve (approx. 6.4 mm [0.25 in.]).

Sediment*—Solid material, both mineral and organic, that is in suspension, is being transported, or has been moved from its site of origin by air, water, gravity, or ice and has come to rest on the earth's surface either above or below sea level.

Sedimentation—The erosion, entrainment, transport, deposition, and compaction of detached soil particles (sometimes used to refer only to the deposition phase of sedimentation).

Sediment Load*—The quantity of sediment, measured in dry weight or by volume, transported through a stream cross section in a given time.

Sheet Flow*—Water, usually storm runoff, flowing in a thin layer over the ground surface.

Side Slope (Engineering)*—The slope of the sides of a canal, dam, or embankment. It is customary to name the horizontal distance first, as 1.5 to 1, or frequently, 1 1/2 : 1, meaning a horizontal distance of 1.5 ft to 1 ft vertical or 1.5 m horizontal to 1 m vertical.

Silt*—1. (Agronomy) A soil separate consisting of particles between 0.05 and 0.002 mm (0.002 in. and 0.00008 in.) in equivalent diameter. 2. A soil textural class. 3. (Engineering) According to the Unified Soil Classification System, a fine-grained soil (more than 50% passing the No. 200 sieve) that has a low plasticity index in relation to the liquid limit.

Sod*—A piece of earth containing grass plants with their matted roots. Turf.

Soil*—1. (Agronomy) the unconsolidated mineral and organic material on the immediate surface of the earth that serves as a natural medium for the growth of land plants. 2. (Engineering) Earth and rock particles resulting from the physical and chemical disintegration of rocks, which may or may not contain organic matter. It includes fine material (silts and clays), sand, and gravel.

Specific Energy*—The average energy per unit weight of water at a channel section as expressed with respect to the channel bottom.

Spillway*—An outfall structure in which the water drops over a vertical wall onto an apron at a lower elevation.

Spreader (Hydraulics)*—A device for distributing water uniformly in or from a channel.

Stabilization*—Providing adequate measures, vegetative and/or structural, that will prevent erosion from occurring.

Stabilized Area*—An area sufficiently covered by erosion-resistant material, such as a heavy cover of grass, or paving by asphalt, concrete, or stone, in order that erosion of the underlying soil does not occur.

Stable (Stream or Channel)—The condition of a stream, channel, or other watercourse that does not change in plan form, cross section, or bed profile during a particular period of time.

Stage (Hydraulics)*—The variable water surface or the water surface elevation above any chosen datum.

Stormwater—Discharges defined as "stormwater runoff, snow melt runoff, and surface runoff and drainage." To be subject to NPDES, stormwater must be in a discernible conveyance (i.e., not sheet flow). Agricultural return flows and runoff are exempt.

Structural (Soil)*—The combination or arrangement of primary soil particles into secondary particles, units or peds. Of note, dune sand is structureless.

Subcritical*—Flow at velocities less than critical velocity.

Temporary Seeding*—A seeding that is made to provide temporary cover for the soil while waiting for further construction or other activity to take place.

Texture (Soil)*—The relative proportions of various soil separates in a soil material.

Time of Concentration*—Time required for water to flow from the most remote point of a watershed, in a hydraulic sense, to the outlet.

TMDL Program —Total Maximum Daily Loads (TMDLs) are waste load allocations on a watershed basis for both point and nonpoint discharges based on a requirement of the Clean Water Act. In this context, urban runoff has been cited as a source of pollutants that are impairing water bodies. EPA has listed "water quality limited water bodies" as those water bodies that are not achieving their beneficial uses. A TMDL defines the amount of a particular pollutant that a water body can absorb on a daily basis without violating water quality standards. States are responsible for implementing the TMDL process by identifying water quality limited waters, establishing priority waters and watersheds, and developing the TMDLs. Once the TMDL is determined, the regulatory agency allocates a portion of the pollutant load to each source within a particular watershed. This assignment is referred to "waste load allocations" and "load allocations."

Topsoil*—Fertile or desirable soil material used to top dress road banks, subsoils, parent material, and other areas and uses.

Trap Efficiency*—The capability of a reservoir to trap sediment. The ratio of sediment trapped to the sediment delivered, usually expressed in percent.

Trash Rack*—Grill, grate, or other device at the intake of a channel, pipe, drain, or spillway for the purpose of preventing oversized debris from entering the structure.

Uniform Flow*—A state of steady flow when the mean velocity and cross-sectional area are equal at all sections of a reach.

Universal Soil Loss Equation (USLE)—An empirical erosion equation developed by the USDA to estimate soil loss by raindrop impact and surface runoff.

$$A = RKLSCP$$

where A is average annual soil loss in tons per acre per year, R is rainfall factor, K is soil erodibility factor, L is length of slope, S is percent of slope, C is cropping and management factor, and P is conservation practice factor.

The RUSLE (Revised Universal Soil Loss Equation) is a computerized version of USLE with improvements in factor estimates. See http://www.ars.usda.gov/Research/docs.htm?docid=18095

Uplift (Hydraulics)*—The upward force of water on the base or underside of a structure.

Waters of the United States—For purposes of the Clean Water Act 33 U.S.C. 1251 et seq. and its implementation regulations, subject to exclusions in paragraph (o)(2) of the section, the term "Waters of the United States" is defined as the following [Clean Water Rule published on June 29, 2015, in the Federal Register (80 FR 37054)]:

(i) All waters which are currently used, were used in the past, or may be susceptible to use in interstate or foreign commerce, including all waters which are subject to the ebb and flow of the tide;

(ii) All interstate waters, including interstate wetlands;

(iii) The territorial seas;

(iv) All impoundment of waters otherwise identified as waters of the United States under this section;

(v) All tributaries, as defined in paragraph (o)(3)(iii) of the section, of waters identified in paragraphs (o)(1)(i) through (iii) of the section;

(vi) All waters adjacent to a water identified in paragraphs (o)(1)(i) through (v) of the section, including wetlands, ponds, lakes, oxbows, impoundments, and similar waters;

(vii) All waters in paragraphs (o)(1)(vii)(A) through (E) of the section where they are determined, on a case-specific basis, to have a significant nexus to a water identified in paragraphs (o)(1)(i) through (iii) of the section. The waters identified in each of the paragraphs (o)(1)(vii)(A) through (E) of the section are similarly situated and shall be combined, for purposes of a significant nexus analysis, in the watershed that drains to the nearest water identified in paragraphs (o)(1)(i) through (iii) of the section. Waters identified in this paragraph shall not be combined with waters identified in paragraph (o)(1)(vi) of the section when performing a significant nexus analysis. If waters identified in this paragraph are also an adjacent water under paragraph (o)(1)(vi), they are an adjacent water and no case-specific significant nexus analysis is required.

APPENDIX E
REFERENCES

AISWCD (Association of Illinois Soil and Water Conservation Districts). (2013). "Illinois urban manual: An erosion and sediment control best management practice manual." <http://www.aiswcd.org/illinois-urban-manual/> (2/11/2016).

ASCE/Environmental and Water Resources Institute (EWRI). (2003). "Regulated riparian model water code." *ASCE/EWRI 40-03*, Reston, VA.

ASCE/EWRI (Environmental and Water Resources Institute). (2009). "Comprehensive transboundary water quality management agreement." *ASCE 33-09*, Reston, VA.

ASTM. (2007). "Standard test method for deterioration of geotextiles by exposure to light, moisture and heat in a xenon arc type apparatus." *ASTM D4355*, West Conshohocken, PA.

ASTM. (2008). "Standard test method for grab breaking load and elongation of geotextiles." *ASTM D4632*, West Conshohocken, PA.

Barrett, M. E., Malina, J. F., Jr., and Charbeneau, R. J. (1998). "An evaluation of geotextiles for temporary sediment control." *Water Environ. Res.*, 70(3), 283–290.

Beecham, T. (2010). "Better borders." *Erosion Contr. J.*, 17(7), 34.

Clark, S. E., Allison, A. A., and Sitler, R. A. (2009). "Geographic variability of rainfall: Impact of rainfall erosivity estimation and construction site erosion control design." *J. Irrig. Drain. Eng.*, 135(4), 474–479.

Claytor, R. (1997). "Practical tips for construction site phasing." *Watershed Prot. Tech.*, 2(3), 413–417.

COMAR (Code of Maryland Regulations). (1974). "Training and certification program." <http://www.dsd.state.md.us/comarhtml/26.17.01.06.htm>; <http://www.dsd.state.md.us/COMAR/searchall.aspx> (Apr. 25, 2017).

Dane County. (2007). "Dane county erosion control and stormwater management manual." <https://olw-wrd.countyofdane.com/documents/PDFs> (4/25/2017).

Daniel, T. C., McGuire, P. E., Stoffel, D., and Miller, B. (1979). "Sediment and nutrient yield from residential construction sites." *J. Environ. Qual.*, 8(3), 304–308.

Dillaha, T. A., Reneau, R. B., Mostaghimi, S., and Lee, D. (1989). "Vegetative filter strips for agricultural nonpoint source pollution control." *Trans. Am. Soc. Agric. Eng.*, 32(2), 513–519.

District of Columbia. (2003). "Standards and specifications for soil erosion and sediment control." <https://doee.dc.gov/soil-erosion-and-sediment-control-handbook> (Jun. 8, 2017).

DNREC (Delaware Natural Resources and Environmental Control). (2013). *Delaware erosion and sediment control handbook*, Delaware Sediment and Stormwater Program, Division of Soil and Water Conservation, Dover, DE.

Downs, S. C., and Appel, D. H. (1986). "Progress report on the effects of highway construction on suspended-sediment discharge in the Coal River and Trace Fork, WV." <https://pubs.usgs.gov/wri/1984/4275/report.pdf> (9/18/2015).

Evans, P. (2004). "Front end support for your regulatory program, Delaware's comprehensive training and education program spreads the responsibility for construction site compliance." *Stormwater*, 5(8).

Everest, F. H., Beschta, R. L., Scrivener, J. C., Koski, K. V., Sedell, J. R., and Cederholm, C. J. (1987). "Fine sediment and salmonid production: A paradox." E. O. Salo and T. W. Cundy, eds. *Streamside management: Forest and fishery interactions*, Univ. of Washington, College of Forest Resources, Seattle.

FAO (Food and Agriculture Organization). (1996). "Control of water pollution from agriculture." <http://www.fao.org/docrep/w2598e/w2598e00.HTM> (2/10/2016).

Federal Register. (2015). "Clean water rule: Definition of 'waters of the United States' rules." *80 FR 37054*, Office of the Federal Register, Washington, DC.

Fennessey, L. A. J., and Jarrett, A. R. (1994). "The dirt in a hole: A review of sedimentation basins for urban areas and construction sites." *J. Soil Water Conserv.*, 49(4), 317–323.

Fennessey, L. A. J., and Jarrett, A. R. (1997). "Influence of principal spillway geometry and permanent pool depth on sediment retention of sedimentation basins." *Trans. ASAE*, 40(1), 53–59.

Fox, H. L. (1974a). "Effects of urbanization on the Patuxent River, with special emphasis on sediment transport, storage, and migration." Ph.D. dissertation, Johns Hopkins Univ., Baltimore.

Fox, H. L. (1974b). "The urbanizing river: A case study in the Maryland piedmont." *Geomorphology and engineering*, D. R. Coates, ed., Hutchinson and Ross, Stroudsburg, PA.

Haan, C. T., Barfield, B. J., and Hayes, J. C. (1994). *Design hydrology and sedimentology for small catchments*, Academic Press, San Diego.

Hainly, R. A. (1980). "The effects of highway construction on sediment discharge into Blockhouse Creek and Steam Valley Run, PA." *WRIR 80-68*, U.S. Geological Survey, Reston. VA.

IECA (International Erosion Control Association). (2001). *Workbook: How to select, install and inspect construction site erosion and sediment control BMPS for NPDES storm water permit compliance*, Steamboat Springs, CO.

Kelsey, C. (2010). "Large-scale testing for erosion and sediment control." *CE News*, Apr. 30.

Klein, S. (1979). "Urbanization and stream quality impairment." <https://www.fws.gov/southwest/es/Documents/R2ES/LitCited/4TX_Sal/Klein_1979_urbanization_and_stream_quality.pdf> (Apr. 25, 2017).

Krause, T., Stansbury, J., Dvorak, B., Admiraal, D., and Young, A. (2008). "Development of a scoring system for construction-site BMP assessment." *Stormwater*, 9(1), 48.

MacDonald, L. H., Smart, A. W., and Wissmar, R. C. (1991). "Monitoring guidelines to evaluate effects of forestry activities on

streams in the Pacific Northwest and Alaska." *EPA 910/9-91-001*, U.S. Environmental Protection Agency, Seattle.
MDE (Maryland Department of the Environment). (1994a). *Maryland standards and specifications for soil erosion and sediment control*, Water Management Administration, Soil Conservation Service, and State Soil Conservation Committee, Baltimore.
MDE. (1994b). "Reserve slope bench." F-19-3.
MDE. (1994c). "Stabilized construction entrance." F-17-3.
MDE. (1994d). "Silt fence." E-15-3.
MDE. (1994e). "Super silt fence." H-26-3.
MDE. (1994f). "Standard inlet protection." E-16-5.
MDE. (1994g). "Pipe outlet sediment trap." C-9-7.
MDE. (1994h). "Stone outlet sediment trap." C-9-10.
MDE. (1994i). "Rip-rap outlet sediment trap." C-9-13.
MDE. (1994j). "Rip-rap/stone outlet sediment trap." C-9-16.
MDE. (1994k). "Earth dike." A-1-6.
MDE. (1994l). "Temporary swale." A-2-4.
MDE. (1994m). "Perimeter dike/swale." A-3-3.
MDE. (1994n). "Pipe slope drain." B-5-4.
MDE. (1994o). "Rip-rap inflow protection." B-6-2.
MDE. (1994p). "Gabion inflow protection." B-7-2.
MDE. (1994q). "Stone check dam." B-8-3.
MDE. (1994r). "Tree protection." G-23-3.
MDE. (2011). "Standards and specifications for soil erosion and sediment control." <http://www.mde.maryland.gov/programs/Water/StormwaterManagementProgram/SoilErosionandSedimentControl/Documents/2011%20MD%20Standard%20and%20Specifications%20for%20Soil%20Erosion%20and%20Sediment%20Control.pdf> (2/11/2016).
Meehan, W. R., ed. (1991). "Influences of forest and rangeland management on salmonid fishes and their habitats." *Special Publication No. 19*, American Fisheries Society, Bethesda, MD.
Megahan, W. F. (1983). "Hydrologic effects of clearcutting and wildfire on steep granitic slopes in Idaho." *Water Resourc. Res.*, 19(3), 811–819.
Megahan, W. F. (1986). "Recent studies on erosion and its control on forest lands in the United States." F. Richard, ed., *Range basin sediment delivery, Albuquerque, NM*, Wallingford, Oxon, U.K., 178–189.
Megahan, W. F., and Kidd, W. J. (1972). "Effects of logging roads on sediment production rates in the Idaho batholith. *Research Paper INT-123*, USDA Forest Service, Intermountain Forest and Range Experiment Station, Ogden, UT.
Meyer, L. L., Wischmeier, W. H., and Daniel, W. H. (1971). "Erosion, runoff, and vegetation of denvded construction sites." *Trans. ASAE*, 14(1), 138–141.
Nelson, J. (1996). "Characterizing erosion processes and sediment yields on construction sites." M. Sc. thesis, Dept. of Civil and Environmental Engineering, Univ. of Alabama, Birmingham, U.K.
New Castle County. (2010). *Internal forms of the department of land use*, New Castle, DE.
Northcutt, G. (2008). "Protecting construction site water quality." *CE News*, Apr. 42–45.
Novotny, V., and Chesters, G. (1981). *Handbook of nonpoint pollution: Sources and management*, Van Nostrand Reinhold, New York.
Novotny, V., and Chesters, G. (1989) "Delivery of sediment and pollutants from nonpoint sources: A water quality perspective." *J. Soil Water Conserv.*, 44(6), 568–576.
NYSDEC (New York State Department of Environmental Conservation). (2005). "New York State standards and specifications for erosion and sediment control." <http://www.dec.ny.gov/docs/water.pdf> (4/25/2017).
PADEP (Pennsylvania Department of Environmental Protection). (2000). *Erosion and sediment pollution control program manual*, Commonwealth of Pennsylvania, Bureau of Watershed Management, Harrisburg, PA.
PADEP. (2012). *Erosion and sediment pollution control program manual*, Commonwealth of Pennsylvania, Bureau of Waterways Engineering and Wetlands, Harrisburg, PA.
Paterson, R., Luger, M., Burby, R., Kaiser, E., Malcom, H., and Beard, A. (1993). "Costs and benefits of urban erosion and sediment control—The North Carolina experience." *Environ. Manage.*, 17(2), 167–178.
Pitt, R., Clark, S. E., and Lake, D. (2007). *Construction site erosion and sediment controls: Planning, design, and performance*, DEStech Publications, Lancaster, PA.
Rafter, D. (2004). "Getting creative with geosynthetics." *Forester Daily News*, Jul. 1.
Schueler, T. (1995). *Site planning for urban stream protection*, Metropolitan Washington Council of Governments, Washington, DC.
USDA (United States Department of Agriculture). (1978). "Predicting rainfall erosion losses: A guide to conservation planning." <https://naldc.nal.usda.gov/naldc/download.xhtml?id=CAT79706928&content=pdf> (Jun. 8, 2017).
USDA. (2016). "Revised universal loss equation 2: Overview of RUSLE2." <https://www.ars.usda-gov/southeast-area-oxford-ms/national-sedimentation-laboratory/watershed-physical-processes-research/rusle2/revised-universal-soil-loss-equation-2-overview-of-rusle2/> (4/25/2017).
USDA/NRCS (United States Department of Agriculture/Natural Resources Conservation Service). (1986). "Urban hydrology for small watersheds." <https://www.nrcs.usda.gov/Internet/FSEDOCUMENTS/stelprdb1044171.pdf> (4/20/2017).
USDA/NRCS. (2001a). "National engineering handbook, part 650: Engineering field handbook, water management (drainage)." <http://directives.sc.egov.usda.gov/viewerFS.aspx?hid=21429> (9/18/2017).
USDA/NRCS. (2001b). "Revised universal soil loss equation, version 2, handbook." <http://www.nrcs.usda.gov/Internet/FSE_DOCUMENTS/nrcs144p2_025079.pdf> (9/18/2017).
USDA/NRCS. (2009). "Conservation practice standard." <https://www.nrcs.usda.gov/Internet/FSE_DOCUMENTS/nrcs143_026468.pdf> (4/15/2016).
USEPA (U. S. Environmental Protection Agency). (1990). "Phase I Stormwater rule." *40CFR122.26*, Washington, DC.
USEPA. (1992). *Managing nonpoint source pollution: Final report to Congress in Section 319 of the Clean Water Act (1989)*, Office of Water, Washington, DC.
USEPA. (1993). *Watershed protection: Catalog of federal programs*, Office of Water, Washington, DC.
USEPA. (1999). "Phase II Stormwater rule." *40CFR122.26*, Office of Water, Washington, DC.
USEPA. (2001). "Stormwater Phase II rule: Construction rainfall erosivity waiver." *USEPA 833-F-00-014*, Office of Water, Washington, DC.
USEPA. (2006). *Model ordinances to protect local resources, model ordinances language, erosion and sediment control*, Washington, DC.
USEPA. (2012). "Fact sheet: EPA issues 2012 general permit for stormwater discharges from construction activities." <https://www.epa.gov/sites/production/files/2015-10/documents/cgp2012shortfactsheet.pdf> (9/18/2015).
USEPA. (2015). "Erosion and sediment model ordinance." <https://www.epa.gov/sites/production/files/2015-12/documents/e-s_model_ordinancel.pdf> (Jun. 8, 2017).

VDEQ (Virginia Department of Environmental Quality). (1992a). *Virginia erosion and sediment control handbook*, 3rd Ed., Richmond, VA.

VDEQ. (1992b). "Typical orientation of treatment–1, soil stabilization blanket." <http://www.deq.virginia.gov/Portals/0/DEQ/Water/StormwaterManagement/Erosion_Sediment_Control_Handbook/Chapter%203%20-%203.36.pdf> (2/11/2016).

VDEQ. (1992c). "Typical treatment–2, soil stabilization matting slope installation." <http://www.deq.virginia.gov/Portals/0/DEQ/Water/StormwaterManagement/Erosion_Sediment_Control_Handbook/Chapter%203%20-%203.36.pdf> (Feb. 2016).

VDEQ. (1992d). "Silt fence culvert inlet protection." <http://www.deq.virginia.gov/Portals/0/DEQ/Water/StormwaterManagement/Erosion_Sediment_Control_Handbook/Chapter%203%20-%203.08.pdf> (Feb. 2016).

VDEQ. (1992e). "Stone culvert inlet protection with sediment trap." <http://www.deq.virginia.gov/Portals/0/DEQ/Water/StormwaterManagement/Erosion_Sediment_Control_Handbook/Chapter%203%20-%203.08.pdf> (Feb. 2016).

VDEQ. (1992f). "Paved flume." <http://www.deq.virginia.gov/Portals/0/DEQ/Water/StormwaterManagement/Erosion_Sediment_Control_Handbook/Chapter%203%20-%203.16.pdf> (Feb. 2016).

VDEQ. (1992g). "Typical waterway cross sections." <http://www.deq.virginia.gov/Portals/0/DEQ/Water/StormwaterManagement/Erosion_Sediment_Control_Handbook/Chapter%203%20-%203.17.pdf> (Feb. 2016).

VDEQ. (1992h). "Portable settlement tank." <http://www.deq.virginia.gov/Portals/0/DEQ/Water/StormwaterManagement/Erosion_Sediment_Control_Handbook/Chapter%203%20-%203.26.pdf> (Feb. 2016).

VDEQ. (1992i). "Dewatering basins." <http://www.deq.virginia.gov/Portals/0/DEQ/Water/StormwaterManagement/Erosion_Sediment_Control_Handbook/Chapter%203%20-%203.26.pdf> (Feb. 2016).

Woelkers, D., and Theisen, M. S. (2003). "BMP selection for regulatory compliance." *Stormwater*, 4(5), 76–89.

Wolman, M. G., and Schick, A. (1967). "Effects of construction on fluvial sediment, urban and suburban areas of Maryland." *Water Res. Res.*, 3(4), 451–464.

Yorke, T. H., and Herb, W. J. (1976). "Urban-area sediment yield effects of construction site conditions and sediment control methods." *Proc., 3rd Federal Inter-Agency Sedimentation Conf.*, Denver, Sedimentation Committee, Water Resources Council, Washington, DC.

Yorke, T. H., and Herb, W. J. (1978). "Effects of urbanization on streamflow and sediment transport in the Rock Creek and Anacostia basins, Montgomery County, MD, 1962–1974." *Professional Paper No. 1003*, U.S. Geological Survey, Reston, VA.

INDEX